Introduction to Relativity

Introduction to Relativity

William D. McGlinn

The Johns Hopkins University Press
Baltimore and London

Printed in the United States of America on acid-free paper

9 8 7 6 5 4 3 2 1

The Johns Hopkins University Press

2715 North Charles Street

Baltimore, Maryland 21218-4363

www.press.jhu.edu

Library of Congress Cataloging-in-Publication Data

McGlinn, William D.
  Introduction to relativity/ William D. McGlinn.
    p. cm.
  Includes bibliographical references and index.
  ISBN 0-8018-7047-X (hc. : alk. paper) – ISBN 0-8018-7053-4
    (pbk. : alk. paper)
    1. Relativity (Physics) I. Title.

QC173.55 .M38 2002
530.11–dc21

                                                    2002070073

A catalog record for this book is available from the British
Library.

To Louise

Contents

Preface

This book provides an introduction to the theory of relativity, both special and general, to be used in a one-term course for undergraduate students, mainly physics and math majors, early in their studies. It is important that students have a good understanding of relativity to appreciate the unifying principles and constraints it brings to both classical and quantum theories. The understanding of such a beautiful subject can bring a great deal of satisfaction and excitement to the student.

The history of both special and general relativity is given short shrift in the book, however. I believe that an initial study of relativity is best done by a straightforward, linear development of the theory, without the twists and turns that are inevitably a part of a theory's history. With such an approach, students can understand and appreciate the structure of the resulting theory more quickly than if the historical path is followed. The historical path has an interest and value of its own, but the illumination of this path is best done by historians of science. Therefore, for example, I have not discussed Mach's principle—the principle that inertial frames are determined by the distribution of mass in the universe. Undoubtedly, Einstein was influenced by Mach, but in the end the answer to the question "Does Einstein's general relativity obey Mach's principle?" is elusive.

The target audience imposes constraints on what material is included and on the level of sophistication, especially mathematical sophistication, employed. I assume the student has had an introductory course in physics, a knowledge of basic calculus, including simple differential equations and partial derivatives, and linear algebra, including vectors and matrices. In addition, it is useful, though not necessary, that students be able to use a symbolic mathematical program such as *Mathematica* or *Maple*. Most physics majors at American universities have the required background at the beginning of their third year, many as much as a year earlier. With such students in mind, I use the traditional tensor index notation and not the coordinate-free notation of modern differential geometry. Though the modern notation gives deeper and quicker insight into the structure of the geometry of space-time, for a first introduction I think the traditional tensor index notation is preferable—and this notation is almost always used in calculations. Also, by the use of symbolic programs students can do calculations that would be some-what prohibitive without them and, thus, develop understanding by way of calculation.

An author of such a textbook faces a choice as to how and when to introduce the required new mathematics, the most important "new," in this case, being tensor analysis (or differential geometry). Should one introduce the mathematics separately so that it stands alone, or should one introduce it within the context of the physics to be described, whereby it can be physically motivated? I do both to some degree. In the study of special relativity, I introduce the concept of the space-time metric and associated tensors within the context of the physics and in analogy with the comparable entities in three-dimensional Euclidean space of which students would be expected to have some understanding. The concepts of a metric and associated tensors under the general coordinate transformations, required in the theory of general relativity, are introduced somewhat abstractly but are quickly tied to the metric of space-time and physical tensors. This path entails a loss of rigor in the discussion. At times, an appeal is made to the reasonableness of a result rather than giving rigorous arguments: it is clear in the context when this is done. One mathematical subject is presented in a "stand alone" manner: in Chapter 8, symmetries of a metric, called *isometries*, are characterized by the existence of Killing vectors. This is a subject one might think is unnecessary in such a book. However, symmetries of the metric are so important in studying dynamically conserved quantities and in the discussion of cosmological models that I feel this introduction to Killing vectors is useful.

The book is roughly divided into two parts. The first part, Chapters 1 through 4, concerns special relativity. General relativity, including a chapter on cosmology, is covered in Chapters 6 through 9. Given the exciting theoretical and observational work being done at the present time, the student should find the study of the cosmology chapter particularly satisfying. Chapter 5 is a reading of Einstein's first attempt at incorporating the equivalence principle into physics, thereby predicting the bending of light rays by the gravitational field of massive bodies, a reading in which the result is viewed as a change in the Minkowskian geometry of space-time. It serves as a physical motivation to changes introduced in his general theory of relativity and is used to relate the coupling constant introduced in Einstein's field equations to Newton's gravitational constant.

Except for Chapter 5, there are exercises associated with each chapter. The number of exercises are few enough so that it is not unreasonable to expect the student to do all of them. The importance of students doing these exercises cannot be emphasized enough. Exercises are particularly important in the study of relativity. The student is susceptible to thinking the material is understood because he or she can "read" the equations. The understanding comes only through serious thought and by working out problems. Furthermore, some important results are contained only in the exercises.

A word about units. A unit of time is introduced so that the speed of light has value one. (In such a set of units, velocities are dimensionless.) This is the only change from standard units, such as MKS units. I do not go "all the way" by defining the Planck units wherein the velocity of light, Newton's gravitational constant G, and Planck's constant h are all chosen to have value one. Such units are particularly useful when studying quantum gravity, a subject we are, most definitely, not considering here.

I am grateful to my late friend, Jim Cushing, and to Steven Shore, who both took the time to read an early version of the notes upon which this book is based and to suggest improvements. Jim's encouragement was instrumental in my decision to put the notes in book form suitable for publication.

Introduction to Relativity

Chapter 1

Foundations of Special Relativity

1.1 Introduction

Before studying Einstein's modification of the scientist's view of space and time as held at the beginning of the last century, we will take a quick look at the theories of physical phenomena that were operative at that time, for the space-time concepts of scientists were shaped by these theories.

At that time, an overall aim of physicists was to give a mechanical explanation of physical phenomena based on Newton's laws of motion. Theories existed for two basic forces: Newton's theory for the gravitational force, and Maxwell's theory for the electromagnetic force.

A relativity principle, recognized by Newton and, before him, by Galileo, applies in Newton's mechanics. With the advent of electromagnetic theory, as incorporated in Maxwell's equations, it seemed that the relativity of Newtonian mechanics could not be the relativity of Maxwell's equations. We study this (Galilean) relativity of Newtonian mechanics to understand its conflict with the (Einsteinian) relativity of Maxwell's equations.

1.2 Kinematics: The Description of Motion

The basic question answered (differently) by Galilean and Einsteinian relativity is, "How do different observers see the same motion?" To address this question, one has to be able to describe the

1

motion of particles quantitatively. A notion prior to motion is that of
an event, that is, a position and a time, for motion is a sequence of
events. To characterize the position of an event, an observer must
choose a coordinate system. One can imagine constructing a lattice
frame of orthogonal (micro) meter sticks so that the position of an
event that occurs at the intersection of three sticks has coordinates
$(ml, nl, ol) = (x, y, z)$, where l is the length of the sticks, and (m, n, o) are
integers. The position $(0, 0, 0)$ is referred to as the origin of the coordinate system and is not a special point, if the space is homogeneous.
We assume one can place the origin of the coordinates at any point
in space without prejudice and can pick the mutually orthogonal axis
pointing in any direction. This is clearly an assumption that space is
homogeneous and isotropic. Of course, an event may not occur
precisely at an intersection, but one can choose the length l as small
as one wishes for the accuracy required. Further, we assume that the
space is Euclidean (flat) and of infinite extent, that there is a distance
between points in the space (x_1, y_1, z_1) and (x_2, y_2, z_2) given by
$d^2 = (x_1 - x_2)^2 + (y_1 - y_2)^2 + (z_1 - z_2)^2$, and that this distance does not
depend on the choice of origin or the direction of the chosen axis.
This Euclidean nature of space is clearly an assumption, one that is
relaxed in Einstein's theory of general relativity. (Imagine trying to
set up a two-dimensional Euclidean lattice on a two-dimensional
sphere.)

The lattice permits one to record the position of an event. How
might the time of an event be characterized? One would like to have
a set of "synchronized" standard clocks at each lattice intersection so
that an event's time measurement can be recorded locally. But how
does one synchronize the clocks? Imagine firing "free" particles from
the origin with a known velocity, in a direction of a lattice junction.
Note that the velocity can be measured locally with nonsynchronized
clocks. Knowledge of the space coordinates of the junction, and thus
the distance from the origin, permits the clock at this junction to be
synchronized with the origin clock, if the velocity of the free particle
is constant and independent of direction. Frames relative to which
free particles, particles that have no forces acting on them, move with
constant velocity are a special class called *inertial frames*. These are
frames in which free particles obey Newton's first law. In discussing
special relativity, we restrict our description of motion to such inertial
frames.

An event is then characterized by a set of space coordinates and a time coordinate. That is, we have four numbers
$(t, x, y, z) = (x_0, x_1, x_2, x_3)$, which we also write as $(t, \mathbf{r})$. The motion of
a particle is described by the position $\mathbf{r}$ of the particle as a func-

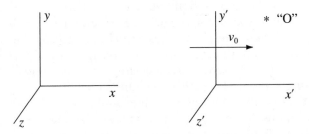

Figure 1.1 Two inertial frames.

tion of the time t—a sequence of events $(t, \mathbf{r}(t))$ with the t ranging continuously over some interval. From this parameterization of the motion, we obtain the instantaneous velocity $\mathbf{v} = d\mathbf{r}/dt$ or, in component form, $(v_x, v_y, v_z) = (dx/dt, dy/dt, dz/dt)$, and the acceleration $\mathbf{a} = d\mathbf{v}/dt = d^2\mathbf{r}/dt^2$.

1.3   Newtonian Mechanics and Galilean Relativity

Newton's force law relates the acceleration $\mathbf{a}$ of a particle with the force F through the equation

$$\mathbf{F} = m\mathbf{a}. \tag{1.1}$$

Here m is a fixed property of the particle, called the (inertial) mass. The value of the force F must be given by some force law if this equation is not to be merely a definition of F in terms of m and $\mathbf{a}$. This law implies that a particle under the influence of no force (a free particle) has $\mathbf{a} = 0$, and thus its velocity $\mathbf{v}$ is constant. The law is valid in an inertial frame.

How are these inertial frames related? What is the relation between the descriptions of the motion of a particle referred to in one inertial frame and another? What is the relation between the force law as viewed in one inertial frame and another? These questions can be answered by a knowledge of the transformation (translation) of events between inertial frames.

Consider two frames: an unprimed inertial frame S and a primed frame S' moving with a constant velocity $\mathbf{v}_0$ relative to the unprimed frame. For simplicity we set up the coordinates of the frames in a particular way. We choose the relative velocity of S' in the x direction and have the origin of the S' frame move along the x axis, the origin of the S frame move along the x' axis, and set the time of the event when the origins agree to be zero for both sets of clocks. This is possible for homogeneous and isotropic space-times. Furthermore, we will assume that we can choose the y' and z' axes such that the the locus of all events with $y = 0$ ($z = 0$) agree with that corresponding to $y'= 0$ ($z'= 0$), with the $z > 0$ ($y > 0$) events agreeing with $z'> 0$ ($y'> 0$). Figure 1.1 depicts a single event "O" with space-time coordinates (t, x, y, z) and (t', x', y', z') as measured in the unprimed and primed frames, respectively. From this figure we see that it is reasonable that events characterized by $x'= 0$ are the set of events characterized by $x - v_0 t = 0$ in terms of unprimed coordinates.

We might expect that the translation of event "O" between the two frames is given by

$$\begin{aligned} t &= t' \\ x' &= x - v_0 t \\ y' &= y \\ z' &= z. \end{aligned}$$

$$(1.2)$$

The first equation assumes that time is universal, that is, that all standard clocks run synchronized. This translation of events is the Galilean transformation for the special coordinates chosen.

This derivation of the Galilean transformation is at best ad hoc. We merely wrote down the transformation equations as one might expect them to be from viewing Figure 1.1. We now tighten up the derivation a bit in a manner that can be generalized to the Einsteinian case and to understand the assumptions hidden in this derivation.[1] For instance, no mention was made about transforming between inertial frames—ones for which a constant velocity observed in one frame implies a constant velocity observed in the transformed frame. It should be reasonable to the reader, and it can be shown, that for this to be so the transformation between the frames must be linear, that is, of the form

$$x'_\mu = \sum_{\nu = 0}^{3} A_{\mu\nu} x_\nu + T_\mu,$$

$$(1.3)$$

where $A_{\mu\nu}$ and T_μ are constants and $x_0 \equiv t$. We assume this to be true. We can choose the space coordinates and the time coordinates so that the event $(0, 0, 0, 0)$, referred to the primed coordinates, is the

---

[1] The derivations of the Galilean and Lorentz transformations that follow are much influenced by those in Rindler (1991).

event $(0, 0, 0, 0)$, referred to the unprimed coordinates, so that $T_\mu = 0$. Eq. (1.3) then becomes

$$x'_\mu = \sum_{v=0}^{3} A_{\mu v} x_v = A_{\mu v} x_v. \qquad (1.4)$$

In the last expression here, we have used the Einstein convention that repeated indexes indicate summation (a convention we will follow hereafter).

What are these constants $A_{\mu v}$ for the transformation between the specially chosen coordinates described above? Again, for our choice of x', y' and z' axes, the events characterized by $x' = 0$ are characterized by $x - v_0 t = 0$. Thus, we have

$$x' = A_{10} t + A_{11} x + A_{12} y + A_{13} z = A_{10} t + A_{11} x = A(x - v_0 t), \qquad (1.5)$$

where A is some constant. If one considered the transformation from the primed coordinates to the unprimed, one would obtain

$$x = A'(x' + v_0 t'), \qquad (1.6)$$

where A' is some constant.

How are A' and A related? If one assumes an equality of the two frames in the sense that no observation can be made that distinguishes one frame as being different or special, then A' must equal A. For instance, consider an event O, say $x_v = (0, L, 0, 0)$, simultaneous with the origin event in the unprimed frame. From Eq. (1.5), for this event $x' = AL$. Similarly, a different event O', say $x'_v = (0, L, 0, 0)$, simultaneous with the origin event in the primed frame, from Eq. (1.6), has $x = A'L$. The assumed equality of the two frames requires $A' = A$ and thus Eq. (1.6) becomes

$$x = A(x' + v_0 t'). \qquad (1.7)$$

Similarly, for our choice of axes, since events characterized by $y' = 0$ are characterized by $y = 0$, the relation

$$y' = A_{20} t + A_{21} x + A_{22} y + A_{23} z = A_{22} y \qquad (1.8)$$

follows. But, if we considered the transformation from the primed to the unprimed frame, we would obtain

$$y = A'_{22} y', \qquad (1.9)$$

and, again with the assumed equality of the frames, $A_{22} = A'_{22}$. With
this, Eqs. (1.8) and (1.9) imply $A_{22}^2 = 1$. Since we chose our axis such
that $y > 0$ corresponds to $y' > 0, A_{22} = A'_{22} = 1$. Similarly, we conclude
that $A_{33} = 1$.

We see that a transformation between the inertial frames (linearity
of transformation) with the axes as chosen and with the assumption
that no inertial frame is special assumes the form

$$x' = A(x - v_0 t)$$
$$x = A(x' + v_0 t')$$
$$y' = y$$
$$z' = z$$
$$t' = A_{00} t + A_{01} x + A_{02} y + A_{03} z. \tag{1.10}$$

Without further assumptions—such as, perhaps, that space is
isotropic—these relations seem to be as far as we can go. The
assumption of a Newtonian universal time implies that $A_{00} = 1$ and
$A_{0i} = 0 \; i = 1, 2, 3$, and these, with the first two relations of Eq. (1.10),
imply $A = 1$. The relations Eq. (1.10) reduce to

$$x' = x - v_0 t$$
$$y' = y$$
$$z' = z$$
$$t' = t, \tag{1.11}$$

which are the Galilean transformation, Eq. (1.2), deduced before.

The transformation of velocity components, defined by
$v_i = dx_i/dt, \; i = 1, 2, 3$ for an arbitrarily moving particle, immediately
obtains

$$v'_x = v_x - v_0$$
$$v'_y = v_y$$
$$v'_z = v_z. \tag{1.12}$$

For the acceleration components, $a_i = dv_i/dt, \; i = 1, 2, 3$, we have

$$a'_x = a_x$$
$$a'_y = a_y$$
$$a'_z = a_z. \tag{1.13}$$

The two different frames record the same acceleration. If $\mathbf{F} = m\mathbf{a}$

is valid in the unprimed frame, then $\mathbf{F'} = m\mathbf{a'}$ is valid in the primed frame if

$$\mathbf{F} = \mathbf{F'}. \tag{1.14}$$

Thus, if Newton's law of motion is valid in one frame, the law is valid in all frames that move with constant velocity with respect to it if (1) the transformation of events is given by a Galilean transformation and if (2) all forces are the same in all such frames. The forces cannot depend on the relative velocity. If the form of the law is not changed by certain coordinate transformations, the law is said to be *invariant* with respect to the transformations considered. Newton's law $\mathbf{F} = m\mathbf{a}$ is invariant with respect to Galilean transformations.

Newton believed in an absolute space or reference frame and that inertial frames were those at rest or in uniform motion with respect to it. However, he recognized the difficulty in discovering which of all inertial frames is the absolute frame because his laws of motion are form invariant under change of frames if the transformation law of events is given by a Galilean transformation and if, under these transformations, the force doesn't change. Such invariance of all physical laws between all inertial frames is the *principle of relativity*. Note that if this principle is valid, then no experiment could be performed that would distinguish one particular frame as being special in any way. (Recall that we used this in the derivation of the Galilean transformation.) The principle of relativity and the assumption of the existence of a universal time give rise to *Galilean relativity*. Force laws, to be consistent with this relativity, must be such that they give an identical force in all inertial frames. Since the vector distance between two simultaneous events is the same in all inertial frames, any law of force between two bodies that depends only on the vector distance between the bodies will satisfy this principle. Newton's gravitational force is such a law. However, a force law that depends upon both velocities of the two bodies (and not just on their relative velocity) would violate this principle.

1.4   Maxwell's Equations and Light Propagation

In 1861 James Clerk Maxwell, born in Edinburgh in 1831, formulated mathematical equations that describe electromagnetic phenomena. These equations predict the existence of electromagnetic waves that travel with a velocity that could be calculated from the theory—calculated in terms of a parameter of the theory that was experimentally determined by measuring the electric force between charges (or the magnetic force between currents). The calculated velocity agreed with that of light as determined by the Danish astronomer Olaus Roemer in 1675 by observing the time lag of the eclipses of Jupiter's moons—a remarkable agreement that could only imply that light is an electromagnetic wave.

In spite of the great successes of Maxwell's equations, most physicists (including Maxwell himself) demanded a mechanical explanation of these waves and, thus, the existence of a medium, called "ether," whose mechanical vibration constituted electromagnetic waves. The speed of light predicted by Maxwell's equations would be the speed in this all-encompassing ether. The principle of Galilean relativity is clearly in trouble. If the ether were a valid concept, it would provide the "absolute" reference frame, thus violating the principle; if Maxwell's equations are valid in all inertial frames, and thus the velocity of light is the same in all inertial frames, the principle of Galilean relativity, which predicts the simple addition of velocities, is invalid.

Irrespective of the existence of the ether, it was crucial to detect the motion of the earth with respect to the frame in which the speed of light has a fixed standard value, independent of the direction of propagation. The most famous attempt to detect this motion was an experiment performed by two Americans, Albert A. Michelson and Edward W. Morley. Michelson first performed the experiment in 1881 and, in collaboration with Morley, again in 1887, with more sensitivity.

Michelson's technique for measuring the motion of the earth through the ether was to split a beam of light in two, send the split beams in two mutually perpendicular directions for approximately equal distances (L in Fig. 1.2), reflect them back to a common point, and measure the difference in travel times of the two. In the figure, the ether is assumed to have a velocity relative to the earth of magnitude v_e, which is assumed to be perpendicular to the direction traveled by the light in one path. One can show (see Exercise 1) that the expected difference in travel time δt of the two beams is

$$\delta t \approx \frac{L}{c}\left(\frac{v_e}{c}\right)^2. \tag{1.15}$$

The difference in arrival times for a reasonable length L is very small —it is proportional to the square of v_e/c. If v_e is taken to be the orbit

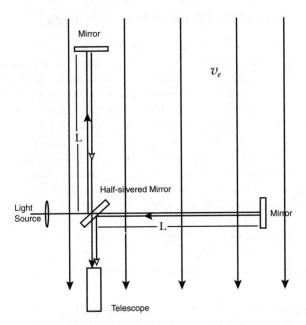

Figure 1.2 Michelson-Morley experimental setup.

speed of the earth around the sun, about 30 km/s, $v_e/c \approx 10^{-4}$, which when squared is a very small number indeed. Michelson did not measure the difference in arrival times directly. Rather, he measured the shift in the interference pattern as the apparatus was slowly rotated. That is, as the apparatus was rotated the arms would interchange the roles of lying parallel and perpendicular to the supposed flow of the ether, thus changing the difference in arrival times of the beams traversing the two arms, resulting in a shift in the fringes of the interference pattern. The total change in the lag time is twice that of Eq. (1.15). A change of one period τ corresponds to a shift of one fringe! The number of fringes n shifted when the apparatus is rotated by 90° is thus

$$n = \frac{2\delta t}{\tau} = \frac{c2\delta t}{\lambda} \approx \frac{2L}{\lambda} \left(\frac{v_e}{c} \right)^2 .$$

Here λ is the wavelength of the light used. The 1887 experiment of Michelson and Morley had $L \approx 11$m, which with $\lambda = 5 \times 10^{-7}$m, a typical wavelength of visible light, yields $n \approx .4$. This is quite detectable— Michelson and Morley had estimated they could detect $n = .01$. However, the experiment gave a null result. No fringe shift was detected.

1.5   Special Relativity: Einsteinian Relativity

It seems that Einstein was not influenced greatly by the Michelson-Morley experiment although he probably knew of the null result. Rather, he doubted the existence of an absolute frame of reference that the Michelson-Morely experiment was attempting to detect. In his later years, Einstein recalled that as a boy of sixteen he wondered how a light wave would appear if one were moving along with it.[2] Galilean relativity would predict a static, nonoscillatory wave. But such a static wave does not satisfy Maxwell's equations *if the parameters that define the theory do not change from one frame to another.* After all, these parameters fix the speed of the wave. Thus, if there is a relativity principle and Maxwell's equations are correct, then Galilean relativity and Newton's mechanics are wrong.

In his remarkable paper of 1905 titled "On the Electrodynamics of Moving Bodies,"[3] Einstein broke with Galilean relativity and its concomitant view of space and time. In this paper he enunciated two postulates:

1. The principle of relativity: All physical laws have the same form (i.e., they are invariant) in all inertial reference frames.

2. The speed of light is the same in all inertial reference frames.

Again, inertial frames are those in which isolated particles, those with no "force" acting on them, move with constant velocity. The first postulate implies that no measurement can be made that distinguishes one frame as being special or different from any other frame.

The second postulate of Einstein, the invariance of the speed of light to all inertial observers, immediately contradicts Galilean transformations of events, since such transformations imply that no speed is invariant. Special relativity is a study of transformations of events, called *Lorentz transformations,* that are consistent with the invariance of the speed of light.

1.5.1   Lorentz Transformation

We now derive the form of the transformation between the two inertial frames depicted in Figure 1.1. Because the speed of light is

---

[2] See *Autobiographical Notes* in Schilpp (1949), p.53. For a view questioning the accuracy of Einstein's recollection, see Bernstein (1973), p.38.

[3] *Annalen der Physik* 17, 891 (1905), reprinted (in English translation) in Einstein et al. (1923).

special—we will see it is the only speed that has the same value for all
inertial observers—it is useful to introduce a time unit that reflects
this special role. The time unit used is the amount of time it takes
light to travel a distance unit. Thus, if we measure distance in meters,
the time unit will be "one meter of light travel time." In such a system
of units, a velocity is dimensionless since time and distance have the
same units. When a velocity is expressed in terms of dimensionless
units we will use the symbol β for velocity. $|\beta| = 1$ for light. Note also
that we usually indicate the time coordinate of an event by "x_0" rather
then "t."

Again, the transformation of events between the two frames is
linear, and we choose the space coordinates and the time coordinates
so that the "origin" events $(0,0,0,0)$ of the two coordinates are the
same event:

$$x'_\mu = A_{\mu\nu} x_\nu. \tag{1.16}$$

As before, the events characterized by $x'_1 = 0$ are characterized by
$x_1 - \beta_r x_0 = 0$, where β_r is the velocity of the primed coordinate with
respect to the unprimed. If one applies the principle of relativity as
before, we again obtain Eqs. (1.10):

$$
\begin{aligned}
x'_1 &= A(x_1 - \beta_r x_0) \\
x_1 &= A(x'_1 + \beta_r x'_0) \\
x'_2 &= x_2 \\
x'_3 &= x_3 \\
x'_0 &= A_{00} x_0 + A_{01} x_1 + A_{02} x_2 + A_{03} x_3,
\end{aligned}
\tag{1.17}
$$

where A is some constant dependent on β_r.

Now, we do not assume a Newtonian universal time $x_0 = x'_0$ as we
did before, and do not obtain $A = 1$. However, consider a sequence of
events of a light pulse traveling in the $+x_1$ (and thus $+x'_1$) direction
whose emission was the origin event $(0,0,0,0)$. These events are characterized (for light $\beta = 1$) by $x_1 = x_0$ and $x'_1 = x'_0$. Using these relations
in the first two equations of Eqs. (1.17), we obtain

$$
\begin{aligned}
x'_1 &= A(x_1 - \beta_r x_0) = A(1 - \beta_r) x_0 = x'_0 \\
x_1 &= A(x'_1 + \beta_r x'_0) = A(1 + \beta_r) x'_0 = x_0.
\end{aligned}
\tag{1.18}
$$

From this it follows that

$$A = (1 - \beta_r^2)^{-1/2}. \tag{1.19}$$

The positive square root is chosen since we assumed the axes were chosen so that a light pulse traveling in the $+x_1$ direction travels in the $+x_1'$ direction. Furthermore, since the first two equations of Eqs. (1.17) are valid for the transformation of any event , we can solve for x_0' in terms of x_0 and x_1, and the last of Eqs. (1.17) becomes

$$x_0' = (1 - \beta_r^2)^{-1/2} (- \beta_r x_1 + x_0). \tag{1.20}$$

The Lorentz transformation of events between these two frames is given by

$$
\begin{aligned}
x_0' &= \gamma \, (- \beta_r x_1 + x_0) \\
x_1' &= \gamma \, (x_1 - \beta_r x_0) \\
x_2' &= x_2 \\
x_3' &= x_3.
\end{aligned}
\tag{1.21}
$$

Here $\gamma = (1 - \beta_r^2)^{-1/2}$, a standard notation. We refer to such a Lorentz transformation, that is Eq. (1.21), as *canonical*. Solving these equations for the x_μ in terms of x_μ', thus obtaining the inverse transformation, one finds

$$
\begin{aligned}
x_0 &= \gamma \, (\beta_r x_1' + x_0') \\
x_1 &= \gamma \, (x_1' + \beta_r x_0') \\
x_2 &= x_2' \\
x_3 &= x_3'.
\end{aligned}
\tag{1.22}
$$

These equations can also be obtained from Eqs. (1.21) by substituting $x_\mu \leftrightarrow x_\mu'$ and $\beta_r \to - \beta_r$.

1.5.2 Lorentz Transformation of Velocities

First note, because the Lorentz transformations are linear, the difference in the space-time coordinates of two events transforms in the same way as the coordinates of a single event, that is,

$$
\begin{aligned}
\Delta x_0' &= \gamma \, (- \beta_r \Delta x_1 + \Delta x_0) \\
\Delta x_1' &= \gamma \, (\Delta x_1 - \beta_r \Delta x_0) \\
\Delta x_2' &= \Delta x_2 \\
\Delta x_3' &= \Delta x_3.
\end{aligned}
\tag{1.23}
$$

Consider then two events, each corresponding to the position of a particle and the time the particle is at the position, with the two events close in time. The above equations give the transformation of the difference in the space-time coordinates of two such events. If one divides the last three of these equations by the first, one easily obtains the transformation equations that relate the components of the velocity of the particle:

$$\beta'_1 = \frac{\Delta x'_1}{\Delta x'_0} = \frac{(\Delta x_1 - \beta_r \Delta x_0)}{(-\beta_r \Delta x_1 + \Delta x_0)} = \frac{\beta_1 - \beta_r}{-\beta_1 \beta_r + 1}$$

$$\beta'_2 = \frac{\Delta x'_2}{\Delta x'_0} = \frac{\Delta x_2 (1 - \beta_r^2)^{1/2}}{(-\beta_r \Delta x_1 + \Delta x_0)} = \frac{\beta_2 (1 - \beta_r^2)^{1/2}}{-\beta_1 \beta_r + 1}$$

$$\beta'_3 = \frac{\Delta x'_3}{\Delta x'_0} = \frac{\Delta x_3 (1 - \beta_r^2)^{1/2}}{(-\beta_r \Delta x_1 + \Delta x_0)} = \frac{\beta_3 (1 - \beta_r^2)^{1/2}}{-\beta_1 \beta_r + 1}. \qquad (1.24)$$

Two important observations should be made about these velocity transformation equations:

1. It is rather easy to argue that if $\beta_1^2 + \beta_2^2 + \beta_3^2 = 1$ then $\beta_1'^2 + \beta_2'^2 + \beta_3'^2 = 1$. Thus, if a particle is moving with a velocity of light in the primed frame, it moves with the velocity of light in the unprimed frame. (See Exercise 2.)

2. If the particle is moving with a constant velocity in one frame, it moves with a constant velocity in the other frame. Thus, if one frame is inertial, the other is as well; these transformations relate inertial frames.

1.5.3   Lorentz Transformation with Arbitrary Relative Velocity

The generalization of the canonical Lorentz transformation Eq. (1.21) for which the relative velocity is in the direction of the x_2 (or x_3) is clear. Thus,

$$x'_0 = \gamma (-\beta_r x_2 + x_0)$$
$$x'_1 = x_1$$
$$x'_2 = \gamma (x_2 - \beta_r x_0)$$
$$x'_3 = x_3. \qquad (1.25)$$

It is useful to know the Lorentz transformation of events between "parallel" frames having arbitrarily directed relative velocity $\boldsymbol{\beta}_r$. In order to derive such transformations, we characterize properties of the canonical transformation with reference to the direction of the relative velocity $\boldsymbol{\beta}$.

First, the equation for x_0' involves only the component of $\mathbf{r}$ in the direction of $\boldsymbol{\beta}$ and can be written in a rotation invariant form as

$$x_0' = \gamma\, (x_0 - \boldsymbol{\beta}_r \cdot \mathbf{r}).$$

Second, the components of the space part of the event that are perpendicular to $\boldsymbol{\beta}_r$ are unchanged. This can be assured by writing

$$\mathbf{r}' = \mathbf{r} + S(\boldsymbol{\beta}_r, \mathbf{r}, x_0)\,\hat{\boldsymbol{\beta}}_r.$$

Here S is a "scalar"— it does not change under a rotation of coordinates.

Finally, for the component of the event in the direction of $\hat{\boldsymbol{\beta}}_r$, $\mathbf{r}_\parallel'$,

$$\mathbf{r}_\parallel' = \gamma \mathbf{r} \cdot \hat{\boldsymbol{\beta}}_r \hat{\boldsymbol{\beta}}_r - \gamma x_0 \boldsymbol{\beta}_r.$$

These last two equations can be combined to yield a vector equation for the transformed space components,

$$\mathbf{r}' = \mathbf{r} + (\gamma - 1)(\mathbf{r} \cdot \hat{\boldsymbol{\beta}}_r)\hat{\boldsymbol{\beta}}_r - \gamma \boldsymbol{\beta}_r x_0, \tag{1.26}$$

which, with the "time" component equation from above,

$$x_0' = \gamma\, (x_0 - \boldsymbol{\beta}_r \cdot \mathbf{r}), \tag{1.27}$$

give a Lorentz transformation of events between "parallel" frames having arbitrarily directed relative velocity $\boldsymbol{\beta}_r$. These frames are "parallel" in the sense that the components of the relative velocity of the primed frame with respect to the unprimed are the negatives of the relative velocity of the unprimed with respect to the primed. This does not imply that the primed axis is parallel to the unprimed axis. Eq. (1.26), a vector expression, written in component form with respect to any (space) coordinate system, gives, with Eq. (1.27), the transformation of the coordinates of events. Note that if $\boldsymbol{\beta}_r$ has components $(\beta_r, 0, 0)$, these expressions reduce to the canonical transformations.

We can easily deduce the transformation of velocities between these "parallel" frames. From Eqs. (1.25) and (1.26), it follows that

$$\boldsymbol{\beta}' = \frac{\Delta \mathbf{r}'}{\Delta x_0'} = \frac{\Delta \mathbf{r} + (\gamma - 1)(\Delta \mathbf{r} \cdot \hat{\boldsymbol{\beta}}_r) \hat{\boldsymbol{\beta}}_r - \gamma \boldsymbol{\beta}_r \Delta x_0}{\gamma (\Delta x_0 - \boldsymbol{\beta}_r \cdot \Delta \mathbf{r})}$$

$$= \frac{\boldsymbol{\beta} + (\gamma - 1) \boldsymbol{\beta} \cdot \hat{\boldsymbol{\beta}}_r \hat{\boldsymbol{\beta}}_r - \gamma \boldsymbol{\beta}_r}{\gamma (1 - \boldsymbol{\beta}_r \cdot \boldsymbol{\beta})}$$

$$= \frac{\gamma (\boldsymbol{\beta} - \boldsymbol{\beta}_r) + (\gamma - 1) \hat{\boldsymbol{\beta}}_r \times (\hat{\boldsymbol{\beta}}_r \times \boldsymbol{\beta})}{\gamma (1 - \boldsymbol{\beta} \cdot \boldsymbol{\beta}_r)}. \tag{1.28}$$

1.6 Exercises

1. Derive Eq. (1.15).
2. Show that if $\beta_1^2 + \beta_2^2 + \beta_3^2 = 1$ in Eq. (1.24) then $\beta_1'^2 + \beta_2'^2 + \beta_3'^2 = 1$.
3. Show that if $\boldsymbol{\beta} \cdot \boldsymbol{\beta} = 1$ in Eq. (1.28), then $\boldsymbol{\beta}' \cdot \boldsymbol{\beta}' = 1$.
4. Show that for the transformation between frames given by Eqs. (1.26) and (1.27), the components of the relative velocity of the primed frame with respect to the unprimed are the negatives of the relative velocity of the unprimed with respect to the primed.
5. In a given intertial frame, two particles are shot out simultaneously from a given point with equal speeds β and in orthogonal directions, say x_1(particle A) and x_2 (particle B). (a) What are the speeds of each projectile relative to the other? (b) What angle does the velocity of B make with the x_1 axis of A as seen by an observer moving with A?

Chapter 2

Geometry of Space-Time

2.1 Introduction

Einstein has changed immeasurably our concept of space-time. The Lorentz transformations imply that the time of an event as measured in one frame depends on both the position and time of the event as measured in a second frame moving with respect to the first. Thus, both the time and the position of a single event are different in different frames. Is there some measure of an event (or difference between two events) that remains invariant (i.e., is the same in all frames)? What characterizes the general transformation between inertial frames?

2.2 Invariant Length for Rotation and Euclidean Transformations

Turning to familiar ground, we study the invariant of the rotation transformations in three space dimensions. Consider two space coordinate systems, relatively stationary and with the same origin, one rotated with respect to the other. The general form of the transformation between such coordinates is given by

$$x_i' = \sum_{j=1}^{3} A_{ij} x_j,$$

(2.1)

such that

$$\sum_{i=1}^{3} x_i'^2 = \sum_{j=1}^{3} x_j^2$$

(2.2)

That is, the squared length of the position vector is invariant under the rotation.

One can consider a more general transformation between coordinate systems for which the origins do not agree, as in

$$x_i' = \sum_{j=1}^{3} A_{ij} x_j + X_i. \tag{2.3}$$

Here X_i are the components of some fixed translation vector. For such a transformation,

$$\Delta x_i' = \sum_{j=1}^{3} A_{ij} \Delta x_j ,$$

and thus

$$\sum_{i=1}^{3} \Delta x_i'^2 = \sum_{j=1}^{3} \Delta x_j^2 \tag{2.4}$$

Such transformations are referred to as *Euclidean transformations*.

2.3 Invariant Interval for Lorentz and Poincaré Transformations

Galilean transformations, including transformations to moving coordinates, are not characterized by a quadratic invariant such as Eq. (2.4). (There is a trivial quadratic invariant for Galilean transformations, namely $\Delta x_0'^2 = \Delta x_0^2$.) However, there is an invariant, comparable to that of Eq. (2.2), of the canonical Lorentz transformations given by Eq. (1.21). The expression $x_0^2 - x_1^2 - x_2^2 - x_3^2$ computed using the primed and unprimed coordinates of an event is the same. Similarly, using the difference of the space-time coordinates of two events, we have

$$\Delta \tau^2 = \Delta x_0^2 - \Delta x_1^2 - \Delta x_2^2 - \Delta x_3^2 = \Delta x_0'^2 - \Delta x_1'^2 - \Delta x_2'^2 - \Delta x_3'^2. \tag{2.5}$$

We call $\Delta \tau^2$ the *invariant interval* (between the two events). The geometry of space-time defined by this invariant interval was introduced by H. Minkowski and is referred to as *Minkowski space*.[1]

We have, then, an invariant of the canonical transformations between the specially related frames. But we can consider the trans-

---

[1] Minkowski's paper, given as an address in 1908, is reprinted (in English translation) in Einstein et al.(1923).

formations between frames whose relative motion is not along both frames' x_1 axis, whose space axis directions do not agree or whose origin events are not the same. We expect that we could effect the most general transformation by performing a sequence of transformations of the canonical type, space rotations, space translations, and time translations. All leave $\Delta \tau^2$ invariant, since space rotations, space translations, and time translations leave Δx_0^2 and $\Delta x_1^2 + \Delta x_2^2 + \Delta x_3^2$ separately invariant. Thus, these more general transformations are of the form

$$x'_\mu = \sum_{\nu = 0}^{3} A_{\mu\nu} x_\nu + X_\mu, \qquad (2.6)$$

such that $\Delta \tau^2$ is invariant. Here X_μ is some constant translation in space-time. Such transformations are referred to as *Poincaré transformations* or *inhomgeneous Lorentz transformations*. The subset of such transformations with $X_\mu = 0$ is called the *homogeneous Lorentz transformations*, or merely the *Lorentz transformations*.

Note that the invariant interval $\Delta \tau^2$ can be positive, negative, or zero in contrast to the rotation invariant distance that is positive for separated points. A positive invariant interval is said to be "timelike," a negative one "spacelike," and a zero invariant interval is said to be "lightlike." Two events corresponding to the position and time of a particle moving with the velocity of light have $\Delta \tau^2 = 0$ in all frames. Conversely, if a particle moves so that $\Delta \tau^2 = 0$, it is moving with the velocity of light. The velocity of light is the same in all frames.

2.4   Space-Time Diagrams

It is sometimes useful, in the study of events and sequences of events, to plot events on a Cartesian space of two or three coordinate axes, with one axis representing time x_0 and the other one or two axes representing, say, x_1 or x_1 and x_2. These Cartesian coordinates are the space and time coordinates of the event for a particular inertial frame. The set of events that are light-like with respect to the origin event, that is, those events which satisfy $\Delta \tau^2 = x_0^2 - x_1^2 - x_3^2 = 0$, form two cones, as depicted in Figure 2.1. These cones are the future and the past light cones of the origin event. Events within these cones are timelike with respect to the origin event; that is, they satisfy $\Delta \tau^2 = x_0^2 - x_1^2 - x_3^2 > 0$, whereas those outside the cones are spacelike, with $\Delta \tau^2 = x_0^2 - x_1^2 - x_3^2 < 0$. The motion of a particle consists of a sequence of events, and events are represented by points in the space-

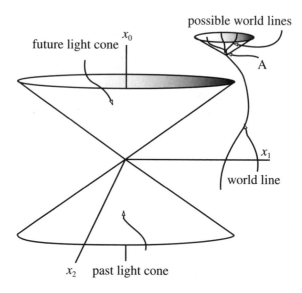

Figure 2.1 Space-time.

time coordinate system. Thus, the motion of a particle is represented by a line, called a *world line*. There is a restriction on the possible world lines of particles. A particle traveling faster than the speed of light has never been observed and, as we will argue, "causality" would seem to dictate that such a velocity is not possible. The magnitude of the velocity of a particle is given by $|\beta| = |\Delta \mathbf{r}/\Delta x_0|$, which is the inverse of the "slope" of the world line—the "slope" of the world line must be greater than (or equal to) one. For any event A that is on the world line of a particle, the ensuing part of the world line must lie in the future light cone of event A (see Fig. 2.1).

Also in the space-time coordinates of the unprimed frame, we can consider the primed coordinates of events, which are related to the unprimed coordinates by the canonical Lorentz transformation Eq. (1.21). The world line of the origin of the primed coordinates, defined by $x_1' = 0$, is given by the locus of events such that $x_1' = 0 = \gamma (x_1 - \beta_r x_0)$ or $x_1 = \beta_r x_0$. This line is depicted in Figure 2.2, labeled as the axis x_0'. It makes an angle $\theta = \tan^{-1} \beta_r$ with the x_0 axis. All lines parallel to this axis are events of constant x_1'. Similarly, the locus of events simultaneous, in the primed frame, with the origin event are characterized by $x_0' = 0 = \gamma (x_0 - \beta_r x_1)$ or $x_0 = \beta_r x_1$. This line is depicted in Figure 2.2, labeled as the axis x_1'. It makes an angle $\theta = \tan^{-1} \beta_r$ with the x_1 axis. All lines parallel to this axis are events of constant x_0', that is, they are loci of events that are simulta-

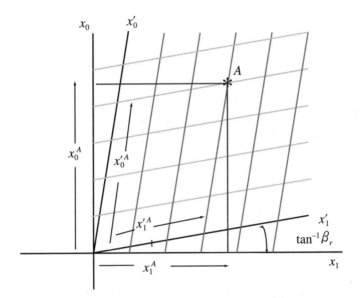

Figure 2.2 Space-time "rotation."

neous in the primed frame. This opposite direction of rotation of
the time axis to that of the space axis reflects the difference between
the geometry of Minkowski space-time and space. In the figure, the
primed and unprimed coordinates of an event A are indicated. A
cautionary note: the scales of the primed and unprimed coordinate
axes are not equal.

2.4.1   Causality

One can argue rather easily that, if two events, A and B, are separated
by a spacelike interval, there are inertial frames for which B occurs
after, before, and simultaneous with A. Choose the A event to be the
origin event and the B event to have vanishing Δx_2 and Δx_3 components in some frame. One can always choose the space axes so that
this is true. Thus, the B event has components $(x_0^B, x_1^B, 0, 0)$ with
$-l^2 = (x_0^B)^2 - (x_1^B)^2 < 0$. Two such events are noted in Figure 2.3. The B
event lies on the hyperbola defined by $-l^2 = x_0^2 - x_1^2 < 0$. The interval's
time component, that is, the time component of the B event, in a
primed frame moving with a velocity β_r in the x_1 direction, is given by

$$x_0^{B\prime} = \gamma(-\beta_r x_1^B + x_0^B). \tag{2.7}$$

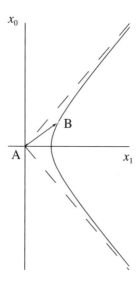

Figure 2.3. Spacelike interval.

The A and B events are simultaneous, in this frame, if $\beta_r = x_0^B / x_1^B$, which has a magnitude less than one since the interval was assumed to be spacelike. In fact, there exists an inertial frame, characterized by some value for β_r, for which $x_0^{B\prime}$ takes on any given value. That is, in viewing the coordinates in Figure 2.3 as "primed" coordinates, there exists a "primed" frame for which the event B has coordinates at any point on the hyperbola. Clearly, two events separated by a spacelike interval cannot be causally related. If, say, A caused (or effected) event B, for some observers B would occur before A and effect would precede cause. This implies that particles cannot have a speed greater than that of light, for with such a particle, event A could be the emission of such a particle and B could be its reception. Event B would then be caused by event A. We conclude that a particle can only proceed into or on its own future light cone.

In a manner similar to the above, one can argue that, if the two events A and B are separated by a timelike interval, say with B in the future light cone of A, as shown in Figure 2.4, then there exists an inertial frame in which the interval has only a time component. That is, the two events occur at the same position. The value of this time component $\Delta\tau = (\Delta\tau^2)^{1/2}$ is called the *elapsed proper time* of the interval. It is the elapsed time of a clock stationary in the frame for which the two events occur at the same position.

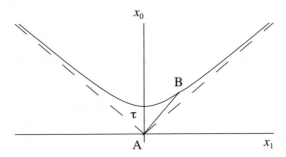

Figure 2.4. Timelike interval.

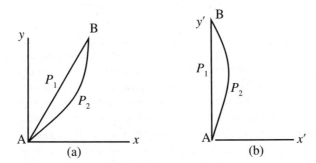

Figure 2.5. Shortest distance.

2.4.2 Longest Elapsed Proper Time between Two Events: The Twin Paradox

The familiar statement, "The shortest distance between two points is a straight line," reflects the geometry of space for which the invariant distance between two nearby points is $(\Delta x^2 + \Delta y^2)^{1/2}$. For consider the two points A and B and paths P_1 and P_2 depicted in Figure 2.5a. P_2 has a longer path length than the straight line path P_1.

This can be shown in a rotated coordinate system (recall that distance is invariant under rotation) in which the path P_1 lies only in the y'–axis as in Figure 2.5b. The total length l_2 of path P_2 is

$$l_2 = \sum (\Delta x'^2 + \Delta y'^2)^{1/2}.$$

But

$$\sum (\Delta x'^2 + \Delta y'^2)^{1/2} \geq \sum (\Delta y'^2)^{1/2} = l_1,$$

where l_1 is the length of path P_1.

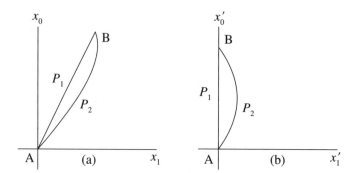

Figure 2.6. Longest proper time.

For a path in space-time—a world line—a comparable statement is: "The *longest* elapsed proper time of a world line between two time-like separated events is that of a straight world line." This reflects the geometry of space-time in which the invariant propertime between two nearby timelike separated events is given by $\Delta\tau = (\Delta x_0^2 - \Delta x_1^2)^{1/2}$. (We consider world lines for which $x_2 = x_3 = 0$, a restriction easily relaxed.) Consider timelike separated events A and B, with coordinates for some inertial frame as shown in Figure 2.6a and two world lines between A and B, one "straight" P_1 and the other not, P_2. We will argue that the total propertime for P_2 is smaller than that of P_1. The path P_1 represents the world line of the origin of an inertial frame moving with constant velocity $\beta = x_1/x_0$, and thus we can imagine these world lines viewed in such a frame, as in Figure 2.6b. Since the total elapsed propertime of a path is the sum of infinitesimal invariants, it can be computed in any inertial coordinate system. The total proper time τ_2 of P_2 is given by

$$\tau_2 = \sum (\Delta x_0'^2 - \Delta x_1'^2)^{1/2} \le \sum \Delta x_0' = x_0'^B = \tau_1. \tag{2.8}$$

Note that we assume that path P_2 moves into its own future light cone. Thus, P_1 has the longest possible elapsed total proper time of any path between A and B. If the paths P_1 and P_2 of Figure 2.6b were the world lines of twins O_1 and O_2, respectively, Eq. (2.8) shows that between the two events A and B corresponding to the crossing of their world lines O_1 has had a longer elapsed time than O_2; O_1 has aged more than O_2. (We assume the biological clock—or any other clock—runs synchronized with the proper time. The principle of relativity demands this.) This is the famous *twin paradox*—the paradox arises, supposedly, because if one considers the process from a coordinate system of O_2

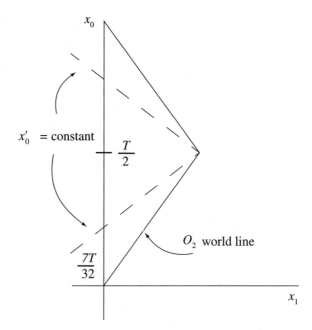

Figure 2.7. Twins' history asymmetry.

one might expect that O_2 should have the longer elapsed time. He would not. There is no paradox—there is no symmetry in the history of the two twins. O_1 stays in a fixed inertial coordinate system; O_2 changes inertial coordinate systems as he undergoes acceleration with respect to O_1.

We can illuminate the asymmetry in the twins' history by considering an example for which the world line of O_2 is particularly simple, as depicted in Figure 2.7. The velocity of O_2 is 3/4 for a time $T/2$, as measured by O_1, after first crossing the world line of O_1. O_2 then changes his inertial frame by changing his velocity to $-3/4$ relative to O_1, who remains in a fixed inertial frame. O_2 returns to intercept O_1's world line at a time T after the first crossing. Two lines of events, simultaneous as observed by O_2, are plotted. One line of events is simultaneous to an event on the world line of O_2 immediately before O_2 turns, the other line is simultaneous to an event on the world line of O_2 immediately after. The elapsed proper time $\Delta\tau_2$ for the two parts of the world line for which O_2 remains in an inertial frame is given by

$$\Delta\tau_2 = \frac{\sqrt{7}}{4}\,(T/2).$$

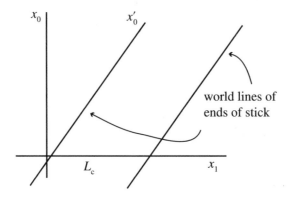

Figure 2.8. Length contraction.

The elapsed proper time $\Delta\tau_1$ of the world line of O_1 as observed by O_2 for each of these periods $\Delta\tau_2$ is $7T/32$. Thus,

$$\Delta\tau_1 = \frac{7}{32}T = \frac{\sqrt{7}}{4}\Delta\tau_2.$$

There is complete symmetry in the behavior of moving clocks as seen from O_1 and O_2 while O_2 remains in a fixed inertial frame. Nevertheless, the total elapsed time for O_2's clock between leaving and returning is $\sqrt{7}\,T/4$, whereas for O_1's it is T.

2.4.3 Length Contraction

We can use the Lorentz transformation of events (or the invariant interval expression) to calculate the length of a moving stick, and we will see that the stick is contracted. It is a matter of asking the correct question. Consider the world lines of the ends of a stick of length L_0 at rest in the primed frame and the two "events" $(0,0,0,0)$ and $(0, L_c, 0, 0)$ on the two world lines and simultaneous in the unprimed frame. (See Fig. 2.8.) L_c is the length "observed" by the unprimed observer. From the Lorentz transformation equations, Eq. (1.21), the event of the "leading" end of the stick, $(0, L_c, 0, 0)$, has primed coordinate $x_1'^0 = L_0 = \gamma L_c$ or $L_c = (1 - \beta_r^2)^{1/2} L_0$; the stick is contracted!

2.4.4 Time Dilation

In a similar manner, we can use the Lorentz transformation of events to calculate what time interval is observed by the unprimed observer

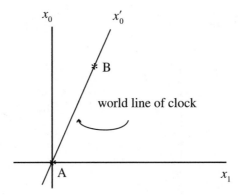

Figure 2.9. Time dilation.

between two events that occur at the same position in the primed frame. The two events can be considered to be two successive ticks of a clock at rest, at $x_1' = 0$, in the primed frame. Consider then two such events, event A, the "origin" event, and B with primed coordinates $(T', 0, 0, 0)$. (See Fig. 2.9.) By use of the Lorentz transformation of events, Eq. (1.22), the event B has unprimed coordinates $x_0 \equiv T = (1 - \beta_r^2)^{-1/2} T' = \gamma T'$ and $x_1 = \beta_r T$. Time is dilated; moving clocks run slowly! We have already seen this time dilation in the twin paradox discussion.

The observation of time dilation, experienced by particles that decay, is made many times each day, thus confirming the prediction of special relativity. As an example, beams of π mesons are produced by bombarding nuclear targets with high-energy protons. These pions leave the target with speeds up to $\beta = .99$. The lifetime of a pion in its rest frame is about 5.4 meters. Thus, if time were not dilated, one would expect the beam to drop in intensity by a factor of $1/e$ each 5.4 meters that it travels. In fact, consistent with time dilation, it decays much more slowly with distance. (See Exercise 6.)

2.4.5 Doppler Shift

Though the speed of light is the same in all inertial coordinate systems, the frequency (and wavelength) changes from one frame to another. This change is referred to as the *Doppler shift*. Space-time diagrams are useful in investigating this shift. Consider a light wave traveling in the $+x$ direction with a period T as observed by the

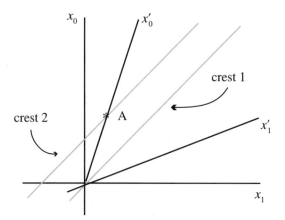

Figure 2.10. Doppler shift.

unprimed observer. What is the period T' in the primed frame moving in the x direction with velocity β_r relative to the unprimed frame?

Figure 2.10 depicts the world lines of successive wave crests, one passing $x_1 = 0$ at $x_0 = 0$ and then the next passing $x_1 = 0$ at $x_0 = T$. Since the crests travel with the velocity of light, their world lines have slope 1. T' is the proper time of event A, with unprimed coordinates (x_0^A, x_1^A) corresponding to crest 2 passing $x_1' = 0$. From the figure, we see that

$$x_0^A = T + x_1^A = T + \beta_r x_0^A .$$

Thus,

$$x_0^A = T/(1 - \beta_r).$$

But

$$T'^2 = (x_0^A)^2 - (x_1^A)^2 = \frac{T^2}{(1 - \beta_r)^2} - \frac{\beta_r^2 T^2}{(1 - \beta_r)^2} = T^2 \frac{(1 + \beta_r)}{(1 - \beta_r)},$$

or

$$T' = T \frac{(1 + \beta_r)^{1/2}}{(1 - \beta_r)^{1/2}} . \tag{2.9}$$

In terms of the frequency, $\nu = 1/T$, we obtain

$$\nu' = \nu \frac{(1 - \beta_r)^{1/2}}{(1 + \beta_r)^{1/2}} .$$

If the primed frame is moving in the positive direction, ν' is smaller than ν. If that frame is moving in the negative direction, ν' is larger than ν.

It is instructive to derive the Doppler shift by considering how a plane light wave, traveling in the $+x$ direction, will look in the two frames. Suppose in the unprimed frame the wave has a wave vector $k = 2\pi/\lambda$ and a frequency $\omega = 2\pi\nu$, with of course $\omega/k = 1$, the velocity of light in our system of units. Let, then, the electric field in the unprimed frame be represented by

$$\mathbf{E} = \mathbf{E}_0 \cos(kx_1 - \omega x_0).$$

The magnetic field of the wave has a similar form,

$$\mathbf{B} = \mathbf{B}_0 \cos(kx_1 - \omega x_0).$$

Though not relevant to our considerations, $\mathbf{E}_0 \cdot \mathbf{B}_0 = 0$ and, for our wave, both $\mathbf{E}_0$ and $\mathbf{B}_0$ lie in the $x_2 - x_3$ plane. The observant reader will note that we are describing a plane-polarized electromagnetic wave. We might expect (see Sec. 4.5) that the electric field, in the primed frame, at a given event is a linear combination of the electric and magnetic fields of the unprimed frame at the *same* event and thus is of the form

$$\mathbf{E}' = \mathbf{E}_0' \cos(kx_1 - \omega x_0),$$

which, when expressed in terms of the primed coordinates of the event, becomes

$$\mathbf{E}' = \mathbf{E}_0' \cos[k\gamma(x_1' + \beta_r x_0') - \omega\gamma(\beta_r x_1' + x_0')]$$
$$= \mathbf{E}_0' \cos[\gamma(k - \omega\beta_r)x_1' - \gamma(-k\beta_r + \omega)x_0'].$$

From this it follows that

$$\omega' = \gamma(-k\beta_r + \omega) = \gamma(1 - \beta_r)\omega,$$

or

$$\nu' = \nu \frac{(1 - \beta_r)^{1/2}}{(1 + \beta_r)^{1/2}}.$$

2.5   Vectors and Scalars

2.5.1   Euclidean Vectors and Scalars

We saw that Newton's equation $\mathbf{F} = m\mathbf{a}$ has an invariant meaning, under Galilean transformations, if the components of $\mathbf{F}$ transform from one inertial frame to another in the same way as the components of $\mathbf{a}$. This statement is of course true if the transformations are restricted to the rotations. The equality of any two vectors has an invariant meaning under rotation if indeed they transform in the same way and we can thus attach an absolute meaning to a vector independent of the particular frame of reference. The components then are the realization of the vector referred to a particular reference frame. With this view, a three-vector is an object whose three components transform, under rotation, from one frame to another with the same transformation as the difference in the coordinates of two positions. Under a rotation, the coordinates of a position transform,

$$x_i' = A_{ij} x_j, \qquad (2.10)$$

in such a way as to leave the distance squared between two positions invariant. We can write this invariance relation as

$$\Delta r^2 = \Delta x_i \delta_{ij} \Delta x_j = \Delta x_i' \delta_{ij} \Delta x_j' = A_{ik} \Delta x_k \delta_{ij} A_{jm} \Delta x_m,$$

where δ_{ij} is the Kronecker delta, defined by

$$\delta_{ij} = \begin{cases} 1, & \text{if } i = j, \\ 0, & \text{if } i \neq j, \end{cases}$$

Since this equality holds for all Δx_k,

$$A_{ik} \delta_{ij} A_{jm} = \delta_{km}, \qquad (2.11)$$

or

$$A_{ik} A_{im} = \delta_{km}. \qquad (2.12)$$

By definition, then, a three-vector $\mathbf{v}$ has components v_i and v_i' related by

$$v_i' = A_{ij} v_j.$$

If one defines the inner product of two vectors $\mathbf{a} \cdot \mathbf{b} = a_i b_i$, then

$$\mathbf{a} \cdot \mathbf{b} = a'_i \, b'_i = A_{ij} \, a_j \, A_{ik} \, b_k = a_j \, \delta_{jk} \, b_k = a_j \, b_j. \tag{2.13}$$

That is, the definition of $\mathbf{a} \cdot \mathbf{b}$ is such that if computed with components of vectors referred to different coordinate axes, the result is unchanged. Such objects are called *three-scalars.*

2.5.2   Lorentzian Vectors and Scalars

If an equation is to be valid in different inertial frames connected by Lorentz transformations, the entities equated must transform in the same way. Recall that the Lorentz transformations—Eq. (2.6) with $X_\mu = 0$—are given by

$$x'_\mu = A_{\mu\alpha} x_\alpha, \tag{2.14}$$

which leaves invariant $\Delta\tau^2$, the invariant interval. We can write the invariant interval of two events as

$$\Delta\tau^2 = \Delta x_\mu \, n_{\mu\nu} \, \Delta x_\nu = \Delta x'_\mu \; n_{\mu\nu} \, \Delta x'_\nu = A_{\mu\nu} \, x_\nu \, n_{\mu\beta} \, A_{\beta\alpha} \, x_\alpha,$$

with

$$n_{\alpha\beta} = \begin{cases} 1, & \text{if } \alpha = \beta = 0, \\ -1, & \text{if } \alpha = \beta = 1, 2, \text{or } 3, \\ 0, & \text{if } \alpha \neq \beta. \end{cases} \tag{2.15}$$

Since this relation is true for any interval Δx_μ, $A_{\mu\alpha}$ satisfies

$$A_{\alpha\rho} \, n_{\alpha\beta} \, A_{\beta\sigma} = n_{\rho\sigma}. \tag{2.16}$$

$n_{\alpha\beta}$ is referred to as the Lorentz or Minkowski metric.[2] Viewed as a matrix, $\mathbf{n}$ is diagonal with diagonal entries $\{1, -1, -1, -1\}$. In contrast, Euclidean space has a metric δ_{ij}.

In order to discuss form-invariant equations, it is useful to define Lorentzian vectors, sometimes called *four-vectors,* as entities that transform under Lorentz transformations as the difference

---

[2] Loosely speaking, a space whose points are characterized by set of continuous variables ζ_μ, on which an infinitesimal "interval" is defined by a symmetric quadratic expression $\Delta^2 = g_{\mu\nu}(\zeta) \, \Delta\zeta_\mu \, \Delta\zeta_\nu$ is called a *metric space* with a metric $g_{\mu\nu}(\zeta)$.

between events Δx_μ. Thus, a four-vector **a** has components a_μ and a'_μ related by

$$a'_\mu = A_{\mu\alpha} a_\alpha. \tag{2.17}$$

Now, defining the "inner" product of two four-vectors by $\mathbf{a} \cdot \mathbf{b} = a_\nu n_{\nu\mu} b_\mu$, we find[3]

$$\mathbf{a} \cdot \mathbf{b} = a'_\nu n_{\nu\mu} b'_\mu = A_{\nu\alpha} a_\alpha n_{\nu\mu} A_{\mu\beta} b_\beta = a_\alpha n_{\alpha\beta} b_\beta.$$

The definition is invariant, that is, it is a four-scalar. Of course, an example of such an inner product is the invariant interval.

The world line of a particle $x_\mu(x_0)$ gives rise to other useful four-vectors. The *four-velocity* U_α is defined by

$$U_\alpha(x_0) \equiv \frac{dx_\alpha}{d\tau}. \tag{2.18}$$

Since dx_α is a four-vector and $d\tau$ is invariant (a scalar), it follows that U_α is a four-vector. We can see how U_α is related to the three-velocity β by noting that

$$d\tau = (dx_0^2 - (dx_1^2 + dx_2^2 + dx_3^2))^{1/2} \tag{2.19}$$

implies

$$\frac{dx_0}{d\tau} = (1 - \beta^2)^{-1/2} = \gamma(\beta), \tag{2.20}$$

and thus

$$U_\alpha(x_0) \equiv \frac{dx_\alpha}{d\tau} = \frac{dx_\alpha}{dx_0}\frac{dx_0}{d\tau} = \gamma(\beta)(1, \boldsymbol{\beta}). \tag{2.21}$$

The relativistic generalization of Newton's force law uses another kinematic variable, the *four-acceleration* A_α defined by

$$A_\alpha \equiv \frac{dU_\alpha}{d\tau}. \tag{2.22}$$

---

[3] We use boldface to indicate four-vectors as we do three-vectors. The type should be clear in context. Actually, we usually indicate vectors by their components and we use Greek letters for four indices and Latin letters for three indices.

2.5.3 The Doppler Shift Revisited

In Section 2.3.5 we derived the Doppler shift for light traveling in the $+x$ direction between two frames in canonical relative motion, that is, for the case where the direction of the light was in the direction of relative motion. It would not have been difficult to generalize that argument to obtain the Doppler shift for any direction of propagation of light. Now, however, we can use the properties of four-vectors and four-scalars to simplify the argument and to illustrate the power of invariance arguments. The same argument as before implies that the phase of the propagating wave is the same in all inertial frames; that is, it is a four-scalar. Let a wave propagating in a direction given by the direction of the three-vector $\mathbf{k}$ be represented by

$$\mathbf{E} = \mathbf{E}_0 \cos(\mathbf{k} \cdot \mathbf{x} - \omega x_0).$$

As noted, the phase of the wave is a scalar:

$$\mathbf{k} \cdot \mathbf{x} - \omega x_0 = \mathbf{k}' \cdot \mathbf{x}' - \omega' x_0'.$$

Since this identity must be true for any event x_μ, $K_\mu = (\omega, \mathbf{k})$ must transform as a four-vector (see Exercise 8), which implies, for a canonical transformation,

$$\begin{aligned}
\omega' &= \gamma(-\beta_r k_1 + \omega) \\
k_1' &= \gamma(k_1 - \beta_r \omega) \\
k_2' &= k_2 \\
k_3' &= k_3.
\end{aligned} \tag{2.23}$$

With $k_1 = k\cos\theta$, $k_1' = k'\cos\theta'$ and, for light, $\omega = k$ and $\omega' = k'$, the first of these equations gives

$$\omega' = \gamma(1 - \beta_r \cos\theta)\,\omega \tag{2.24}$$

or

$$\nu' = \gamma(1 - \beta_r \cos\theta)\,\nu. \tag{2.25}$$

Here $\theta\,(\theta')$ is the angle the direction of propagation of light makes with the $x_1\,(x_1')$ axis. An expression for θ' in terms of θ can be obtained by dividing the second equation by the first equation of Eq. (2.23):

$$\frac{k_1'}{\omega'} = \frac{k'\cos\theta'}{\omega'} = \cos\theta' = \frac{\cos\theta - \beta_r}{1 - \beta_r\cos\theta} \tag{2.26}$$

This direction relationship can also be obtained by transforming the velocity of a pulse of light. (See Exercise 3.)

2.6   Rotation and Lorentz Transformations as Groups

Note that if the transformation equation Eq. (2.10) is written as a matrix equation,

$$\mathbf{x}' = \mathbf{Ax}, \tag{2.27}$$

then Eq. (2.12) written in matrix form is

$$\tilde{\mathbf{A}}\mathbf{A} = \mathbf{I}. \tag{2.28}$$

Any matrix transformation, $\mathbf{A}$, that leaves the distance squared invariant has an inverse given by its transpose, $\tilde{\mathbf{A}}$. $\tilde{\mathbf{A}}$ also leaves the distance squared invariant, that is, $\mathbf{A}\tilde{\mathbf{A}} = \mathbf{I}$. The matrix product $\mathbf{C}$ of two such matrices, $\mathbf{A}$ and $\mathbf{B}$, $\mathbf{C} = \mathbf{BA}$, leaves the distance squared invariant and has as an inverse, $\mathbf{C}^{-1} = \tilde{\mathbf{C}} = \tilde{\mathbf{A}}\tilde{\mathbf{B}}$. Thus, the set of matrices that leaves the distance squared invariant satisfies the group properties,[4] with with the group multiplication defined by matrix multiplication. This group is called the *rotation group.*

Similar to rotation transformations, Lorentz transformations can be realized by matrices. If the transformation equation, Eq. (2.14), is viewed as a matrix equation,

$$\mathbf{x}' = \mathbf{Ax}, \tag{2.29}$$

then Eq. (2.16), in matrix form, becomes

$$\tilde{\mathbf{A}}\mathbf{n}\mathbf{A} = \mathbf{n}. \tag{2.30}$$

From this it follows that $\mathrm{Det}\,\mathbf{A} = \pm 1$, where Det signifies the determinant. Thus, any matrix transformation, $\mathbf{A}$, that leaves $\Delta\tau^2$ invariant,

---

[4] A group $\mathcal{G}$ is a set such that (1) a multiplication that associates with each pair of elements a, b of $\mathcal{G}$ a third element c, written $c = ab$; (2) the multiplication is associative, $(ab)c = a(bc)$; (3) there exists an element e (the identity), such that $ae = ea$ for all $a \in \mathcal{G}$; and (4) for all $a \in \mathcal{G}$ there exists an element a^{-1} such that $aa^{-1} = a^{-1}a = e$.

has an inverse A^{-1}. It is easy to show that $\tilde{A}^{-1} n A^{-1} = n$, that is, A^{-1} leaves $\Delta \tau^2$ invariant. The matrix product C of two such matrices, A and B, $C = BA$, leaves the $\Delta \tau$ invariant. Thus, the set of matrices that leave $\Delta \tau$ invariant satisfies the group properties with the group multiplication defined by matrix multiplication. The group is called the *Lorentz group*.

2.7 Exercises

1. A stick of length L' at rest in the primed frame makes an angle θ' with the x' axis, as shown in Figure 2.11. What angle does the stick make with the x axis in the unprimed frame?

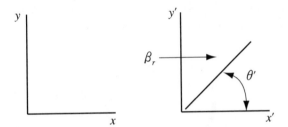

Figure 2.11. Moving stick.

2. (A generalization of Exercise 1) A stick of "apparent" length L' in the primed frame moves with a velocity $\boldsymbol{\beta}' = \beta_x' \, \mathbf{i} + \beta_y' \, \mathbf{j}$, making an angle θ' with the x' axis, as shown in Figure 2.12. (a) What angle does it make with the x axis? (b) With what velocity is it moving in the unprimed frame? (c) Does your answer to (a) reduce to that of Exercise 1 if $\beta' = 0$? (d) If $\beta_x' = 0$ and $\theta' = 0$ what is θ?

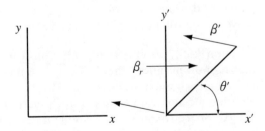

Figure 2.12. Another moving stick.

3. A pulse of light is emitted at an angle θ' with respect to the x' axis in the primed frame. What angle does the direction of the light

make with the x axis in the unprimed frame? Why is this angle different from θ in Exercise 1?

4. Argue that, if two events A and B are separated by a timelike interval, $\Delta\tau^2 > 0$, then there exists an inertial frame in which the interval AB has only a time component.

5. Express the four-acceleration A of a particle in terms of the three-acceleration $\mathbf{a}$ and the three-velocity $\boldsymbol{\beta}$. What is the value of the inner product $A \cdot U = A_\alpha \eta_{\alpha\beta} U_\beta$ in the rest frame of the particle (and thus in any frame, since the inner product is a scalar)?

6. Two π^+ mesons are created together. One moving with a velocity β is immediately inserted into a storage ring of radius R, as illustrated in Figure 2.13. The second is created at rest. After one time around the ring, thus rejoining its twin pion, what is the age of the

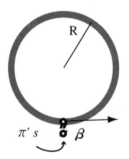

Figure 2.13. Twin pions.

moving pion? What is the age of twin pion at rest at this time? The pions have a lifetime of about 5.4 meters. If $\beta = .98$ and $R = 30m$, what is the probability that the moving pion survives one time around the ring? And what is the probability that the stationary pion is still "alive" when the moving pion returns?

7. As shown in Section (2.6), Lorentz transformations can be implemented by matrix multiplication, that is, by a 4×4 matrix acting on a 4-column vector whose components are x_μ. (a) What is the matrix representing a canonical Lorentz transformation with relative velocity β_r? (b) Space rotations are Lorentz transformations— they leave $d\tau^2$ unchanged. What is the 4×4 matrix that represents rotation about the x_3 axis by an angle θ? (c) What is the matrix that represents the Lorentz transformation that results if one first implements a rotation (b) with $\theta = \theta_0$, followed with a Lorentz transformation (a) and then a rotation (b) with $\theta = -\theta_0$? (d) What

is the result of (c) if $\theta_0 = \pi/2$? Might you have anticipated the result? (It is advisable to use a symbolic program, *Mathematica* or *Maple.*)

8. By use of Eq. (2.30), $\tilde{\mathbf{A}}\mathbf{n}\mathbf{A} = \mathbf{n}$, show that if $\mathbf{Knx} = \mathbf{K'nx'}$ for all four-vectors $\mathbf{x}$ then $\mathbf{K}$ must transform as a four-vector.

9. Show that Eqs. (1.26) and (1.27) imply

$$x_0'^2 - \mathbf{r}' \cdot \mathbf{r}' = x_0^2 - \mathbf{r} \cdot \mathbf{r}.$$

Chapter 3

Relativistic Dynamics

3.1 Introduction

Newtonian mechanics satisfies the Galilean but not the Einsteinian relativity principle. Clearly, then, the dynamical laws of Newton must be modified so that they are consistent with the Einsteinian relativity principle. First, we show how an important law of Newtonian mechanics—the conservation of momentum—is modified. Recall this law in Newtonian mechanics. If two particles collide, that is, interact, then the law states that the total momentum after the collision is the same as before. In Newtonian mechanics the momentum $\mathbf{p}$ of a particle of mass m, moving with velocity $\mathbf{v}$, is given by

$$\mathbf{p} = m\mathbf{v}. \tag{3.1}$$

That momentum is conserved and that the momentum of a particle is given by Eq. (3.1) follow from Newton's third and second laws of motion. Rather than viewing conservation laws as resulting from a particular form of a theory, one can take a more general view that they arise from assumed symmetries of the physical theory. Thus, the conservation of a vector, the momentum, follows from the homogeneity of space. That is, the theory is invariant under the change of frames by translations. The particular form the momentum takes, for instance, Eq. (3.1), results from invariance of the conservation laws under the transformations between different inertial frames. Since Galilean relativity has a transformation law different than that of Einsteinian relativity, it is not surprising that the form of the momentum for Newtonian mechanics must differ from that required by the special theory of relativity. In the following we assume that a vector

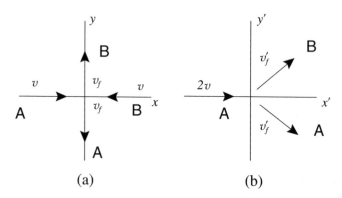

Figure 3.1. (a) Collision in zero momentum frame (b) Collision in lab frame.

quantity associated with the colliding particles is conserved in all inertial frames and deduce the form this vector quantity must take for Galilean relativity and Einsteinian relativity.

We will see, of necessity, that the relativistic momentum of a particle has an associated relativistic energy and that both are inexorably mixed by Lorentz transformations. This gives rise to an energy-momentum geometry, just as there is a space-time geometry as studied in the last chapter.

Finally, in this chapter, the relativistic form of Newton's force law and the relativistic dynamics of the gyroscope will be studied.

3.2   Momentum in Galilean Relativity

Consider the simple case of the collision of two identical particles. We want to associate a vector quantity **p** with the particle moving with constant velocity **v**. From the assumed isotropic property of space, it follows that $\mathbf{p} = f(v^2)\mathbf{v}$; that is, **p** must point in the direction of **v**, the only direction characterized by the motion.

We show that Galilean relativity implies $f(v^2)$ is a constant. The collision we consider is that of two identical particles, A and B, which before the collision move with velocities of magnitude v and rebound at right angles with velocity magnitude v_f, as depicted in Figure 3.1a. For any $f(v^2)$, the total momentum is zero before the collision. Thus, the particles A and B must come off with velocities of the same magnitude but of opposite directions. Now consider the same collision in a frame moving with velocity v in the negative x direction, as illustrated in Figure 3.1b. By the Galilean transformation of velocities, Eqs.

(1.12), in this frame B is at rest and A is moving with velocity $2v$ in the positive x direction before the collision, whereas after the collision

$$\mathbf{v}'^{B}_{f} = v\mathbf{i} + v_f\,\mathbf{j}$$
$$\mathbf{v}'^{A}_{f} = v\mathbf{i} - v_f\,\mathbf{j}.$$

Conservation of the x component of momentum gives

$$f(4v^2)\,2v = f(v^2 + v_f^2)\,v + f(v^2 + v_f^2)\,v.$$

This implies

$$f(4v^2) = f(v^2 + v_f^2),$$

which must be true for any v and v_f. Thus, $f(v^2)$ is a constant m called the *mass of the particle*. The momentum of the particle is, as expected,

$$\mathbf{p} = m\mathbf{v}. \tag{3.2}$$

3.3 Momentum-Energy in Einsteinian Relativity

The expression $\mathbf{p} = m\boldsymbol{\beta}$ for the momentum is valid even in Einsteinian relativity for particles with velocities much smaller than that of light ($\beta \ll 1$).This is true because the Lorentz transformation of velocities is approximated by the Galilean transformations of velocities if the relative velocities of the two frames considered is small compared to that of light. However, if the particle is moving with large velocity, we should expect that $\mathbf{p} = m\boldsymbol{\beta}$ cannot be the expression for the momentum, since transformations of velocities under Lorentz transformations differ from those of Galilean transformations.

We want, then, to deduce the required form of the momentum by considering the collision of two identical particles using the Lorentz transformation between frames. The momentum is to be a three-vector, meaning it transforms as a vector under rotations. As before, the isotropic property of space implies $\mathbf{p} = f(\beta)\,\mathbf{U}$. We use here $\mathbf{U} \equiv \gamma(\beta)\boldsymbol{\beta}$, the space part of the four-velocity, rather then the velocity $\boldsymbol{\beta}$ because of the simple transformation properties of the four-velocity. The space part of the four-velocity does point in the direction of $\boldsymbol{\beta}$. Consider again the collision of the identical particles, first in a frame in which they are moving with equal and opposite velocities of magnitude β_i and rebound at right angles with velocities of magnitude β_f. (See Fig. 3.2a.) The geometry of the collision is the

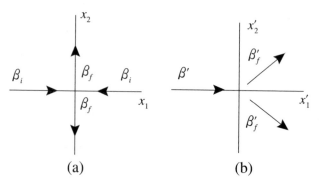

Figure 3.2. Relativistic collision.

same as that considered for the Galilean case. Clearly, the total momentum is zero, independent of the form of $f(\beta)$, both before and after the collision, and it is thus conserved. Now, view the collision in a frame moving, with respect to the first frame, in the negative x_1 direction with a velocity whose magnitude is β_i. (See Fig. 3.2b.) By use of the transformation equations for velocity, Eqs. (1.24), and four-velocity—which of course transforms as a four-vector, Eq.(1.23)—we can calculate the total momentum before and after the collision in the primed frame. For the velocity β' of the incoming particle and the x_1' and x_2' components β_1' and β_2' of the outgoing particles, we obtain (refer to Fig 3.2 b)

$$\beta' = \frac{2\beta_i}{1+\beta_i^2}$$
$$\beta_1' = \beta_i$$
$$\beta_2' = \beta_f(1-\beta_i^2)^{1/2}.$$

$$(3.3)$$

The x_2 component of the total momentum before and after the collision is zero, independent of the function $f(\beta)$. However, the conservation of the x_1' component of the momentum requires the equality

$$f\left(\frac{4\beta_i^2}{(1+\beta_i^2)^2}\right)\left[U_1^i + \beta_i U_0^i\right] = 2f\left(\beta_i^2 + \beta_f^2(1-\beta_i^2)\right)\beta_i U_0^f.$$

$$(3.4)$$

For the case $\beta_i = \beta_f$, an elastic collision, Eq. (3.4) reduces to

$$f\left(\frac{4\beta_i^2}{(1+\beta_i^2)^2}\right) = f\left(\beta_i^2 + \beta_f^2(1-\beta_i^2)\right). \tag{3.5}$$

For this to be true for any β_i, $f(\beta^2)$ must be a constant that is independent of β^2, a scalar property of the particle called its (invariant) mass, m. Thus, by considering an elastic collision, we have determined the relativistic expression for the momentum, an expression that must be valid in inelastic collisions. With f set equal to m, Eq. (3.4) becomes

$$\frac{2\beta_i}{(1-\beta_i^2)^{1/2}} = \frac{2\beta_i}{(1-\beta_f^2)^{1/2}}. \tag{3.6}$$

This implies β_i must equal β_f; the collision *must* be elastic!

The particle's momentum is the mass times the proper time rate of change of position;

$$\mathbf{p} = m\frac{d\mathbf{r}}{d\tau} = m\gamma\boldsymbol{\beta}. \tag{3.7}$$

For small velocities,

$$\mathbf{p} \approx m\frac{d\mathbf{r}}{dx_0} = m\mathbf{v},$$

the Galilean expression for the momentum.

Note that the expression for $\mathbf{p}$ is the space-part of a four-vector. The time component of this four-vector is

$$p_0 = m\frac{dx_0}{d\tau} = \gamma m. \tag{3.8}$$

This four-vector, with components (p_0, p_1, p_2, p_3), is referred to as the *energy-momentum four-vector*, or merely the *momentum four-vector*. If one knows the three-momentum $\mathbf{p}$ and mass m of a particle, and thus knows p_0, in one frame, one may determine, by use of the transformation properties of a four-vector, the value of the four-momentum in any frame. Using this one can easily argue that, if the total three-momentum is to be conserved in all frames, then the total time component of the four-momentum must be conserved, as we determined in our example of the collision of identical particles. This time component p_0 is called the *energy of the particle*.

The momentum four-vector has the components

$$p_\mu = mU_\mu. \tag{3.9}$$

We see that the energy of a particle at rest is equal to its mass m. (Note that the energy has units of mass. If one wishes to express the energy in conventional units, one would multiply by c^2, giving $p_0 = mc^2\gamma$, which for a particle at rest becomes $E = p_0 = mc^2$).

It is reasonable to define the kinetic energy T as the energy a particle has as a result of motion; thus,

$$T = p_0 - m = \gamma m - m,$$

which for low velocities ($\beta \ll 1$) becomes

$$T \approx \frac{m\beta^2}{2},$$

the expression for kinetic energy in Newtonian mechanics.

3.4   The Geometry of the Energy-Momentum Four-Vector

Just as $\Delta\tau^2 = \Delta x_\alpha \eta_{\alpha\beta} \Delta x_\beta$ is invariant under Lorentz transformations, so too is the expression $p_\alpha n_{\alpha\beta} p_\beta = m^2$. The four-momentum for a massive particle is a timelike four-vector in the sense that its associated invariant is greater than zero. If one considers a particle of a mass m, the possible values of its energy-momentum components are restricted by the relation

$$p_0^2 - (p_1^2 + p_2^2 + p_3^2) = m^2. \tag{3.10}$$

Using coordinate axes of p_0, p_1, and p_2 (and setting $p_3 = 0$) we can plot this equation and obtain a hyperbola, called the *mass hyperbola*, pictured in Figure 3.3. The values of the components of four-momentum of a particle of mass m must lie on this hyperbola. For instance, a particle at rest has $p_1 = p_2 = 0$ and $p_0 = m$. Note also that the four-momentum must lie in the "forward" cone since the energy p_0 is greater than zero. There is no mass hyperbola in the "backward" cone.

In the limit of $m \to 0$, the mass hyperbolas approach the cone characterized by $p_0^2 - (p_1^2 + p_2^2) = 0$, a mass hyperbola of zero mass. The expression for the energy $p_0 = m\gamma$ and the momentum $\mathbf{p} = m\gamma\beta$ are both zero if $m = 0$, unless $\beta = 1$, in which case the expressions are indeterminate. However, for very fast moving particles $p_0 \approx |\mathbf{p}|$; for $p_0 = |\mathbf{p}|$ the invariant mass is zero. For these reasons we consider particles that travel with the speed of light to have invariant

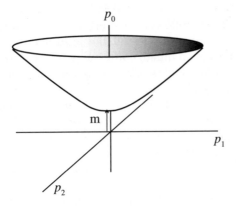

Figure 3.3. Finite mass hyperbola.

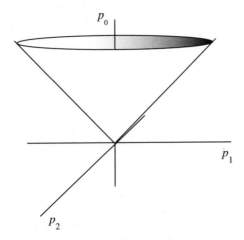

Figure 3.4. Zero mass hyperbola.

mass zero. Their mass hyperbola is the cone depicted in Figure 3.4. Conversely, particles with zero invariant mass travel with the speed of light. One can easily show (see Exercise 1) that, if the four-momenta of two particles are added (i.e., if the corresponding components are added to obtain the components of the sum), the resulting four-momentum is timelike, the invariant being greater than zero, or light like, the invariant being zero. It is lightlike only if the two original four-momentum vectors are themselves lightlike, with their space momenta parallel. It follows that in adding the four-momenta of any number of particles, one always obtains a timelike four-vector (unless, of course, all the particles' four-momenta that are added are lightlike with all three-momentum vectors parallel). This four-momentum has a nonzero invariant mass associated with it. Since this total four-vector is always timelike, it follows that there exists a frame in which the space components vanish. (Recall the argument that there always exits a frame in which a timelike interval is a pure time interval.) The surviving time component is of course the invariant mass of the total four-momentum. Such a frame is referred to as the *center of mass frame*, or *zero-momentum frame*. It is the frame in which the total three-momentum is zero. As in Newtonian physics, it is quite often a useful frame to use in the description of collisions. Of course, knowledge of the four-momenta of the particles before and after collision in one frame implies knowledge of these four-momenta in any frame, since they transform as four-vectors.

It should be noted that the relativistic expression Eq. (3.10), which relates the energy and three-momentum of a particle, is confirmed every day in many laboratories where relativistic collision experiments are performed and analyzed, thus confirming the theory of special relativity.

3.4.1   "Elastic" Collisions

Consider, first, in the zero-momentum (unprimed) frame, the collision of two identical particles, each of mass m, moving along the x_1 axis in opposite directions with velocity magnitude β. Without loss of generality, we can write that after collision the particles come off with velocity components $\pm(\beta_1^f, \beta_2^f, 0)$ with $(\beta_1^f)^2 + (\beta_2^f)^2 = \beta^2$. This latter relation follows from the conservation of energy. Now transform these final momenta to the laboratory (primed) frame (where one particle is at rest before the collision), a frame that moves with speed β with respect to the zero-momentum frame. In this lab frame, one particle rebounds with three-momentum components $(\gamma(m\beta\gamma + m\gamma\beta_1^f), m\gamma\beta_2^f, 0)$ and the other with $(\gamma(m\beta\gamma - m\gamma\beta_1^f), -m\gamma\beta_2^f, 0)$. If θ and ϕ denote the angles that the outgoing particles make with the incident path (i.e., in our case the x_1' axis), we obtain

$$\tan\theta \tan\phi = \frac{\beta_2^f}{(\beta\gamma - \gamma\beta_1^f)}\frac{\beta_2^f}{(\beta\gamma + \gamma\beta_1^f)} = \frac{1}{\gamma^2}.$$

Since $\tan\theta \tan\phi \neq 1$, we see that the angle between the paths of the "rebounding" particles is not $\pi/2$. The particles always rebound in orthogonal directions in elastic Newtonian collisions of identical particles as viewed in the lab frame.

3.4.2 "Inelastic" Collisions

In Newtonian collisions (kinetic) energy need not be conserved. Such a collision is referred to as a inelastic collision, and a particularly simple highly inelastic collision is one in which the particles stick together after the collision occurs. However, in Einsteinian relativity the conservation of three-momentum implies the conservation of the fourth component of the momentum, the energy. What *does* change in a collision in which the particles stick together?

Consider such a collision of two identical particles, each of mass m_0. We can view the collision in the zero-momentum frame in which the particles are moving toward each other with equal speed. (Consider drawing the four-momenta of the two particles in a coordinate system with axes p_0 and p_1.) Since there is only one particle after the collision, the total four-momentum must be the four-momentum of the final particle. In the zero-momentum frame, the four-momentum has only a time component and the final particle is at rest. It has a mass m given by

$$m = p_0 = \gamma m_0 + \gamma m_0 = 2\gamma m_0 > 2m_0.$$

The mass of the a final particle is greater then the sum of the masses of the two original particles.

3.4.3 Particle Production

Another interesting example of a collision is one involving particle production. In contrast to the "inelastic" collision in which a single particle emerges from the collision, in particle production more particles emerge than collide. Consider the case in which an electron e^- collides with a positron e^+, a particle with the same mass m_e as an electron but with the opposite charge, and produces an additional electron and an additional positron. Thus, we are considering the process

$$e^- + e^+ \rightarrow 2e^+ + 2e^-.$$

In the zero-momentum frame, the two particles are traveling in opposite directions with equal speeds β. What is the minimum speed they must have so that the process can occur? Again, the total four-momentum has only a time component p_0 that must have a magnitude of at least $4m_e$, the value if all of the particles are produced at rest. Since the incoming particles each have a time component of momentum equal to $m_e \gamma$, we have $2m_e \gamma = 4m_e$ if the particles are produced at rest. Thus, $\beta = \sqrt{3}/2$ is the minimum speed required to produce the additional electron and positron.

3.5   Relativistic Form of Newton's Force Law

Newton's force law,

$$\mathbf{f} = \frac{d\mathbf{p}}{dx_0}, \tag{3.11}$$

does not behave nicely under Lorentz transformations. Thus, $\mathbf{f}$ of one frame is not equal to $\mathbf{f}'$ of another frame, in contrast to what obtains under Galilean transformations Eq. (1.14). An obvious relativistic generalization of Eq. (3.11) is

$$F_\alpha = \frac{dp_\alpha}{d\tau}. \tag{3.12}$$

F_α is referred to as the *four-force*. The usefulness of such an entity is of course manifest only if the four-force is given by some force law. However, the four-force can be related to the *relativistic* three-force $\mathbf{f}$ *defined* by Eq. (3.11):

$$F_\alpha = \frac{dx_0}{d\tau}\frac{dp_\alpha}{dx_0} = \gamma(\beta)\left(\frac{dp_0}{dx_0}, \frac{dp_i}{dx_0}\right) = \gamma(\beta)\left(\frac{dp_0}{dx_0}, \mathbf{f}\right). \tag{3.13}$$

Furthermore, we have

$$\frac{dp_0}{dx_0} = \frac{d(p^2 + m^2)^{1/2}}{dx_0} = \frac{\mathbf{p}\cdot\mathbf{f}}{p_0} = \boldsymbol{\beta}\cdot\mathbf{f}, \tag{3.14}$$

so that the rate at which the (relativistic) energy is changing in a given frame is equal to the dot product of the velocity and the (relativistic) three-force in that frame. Note also that in the instantaneous rest frame of the particle, Eq. (3.13) reduces to

$$F_\alpha = (0, \mathbf{f}) = \left(0, \frac{d\mathbf{p}}{dx_0}\right). \tag{3.15}$$

3.6   Dynamics of a Gyroscope

In Newtonian mechanics, the rotational dynamics of a rigid body is governed by the equation

$$\tau_{cm} = \frac{d\mathbf{L}_{cm}}{dt}. \qquad (3.16)$$

Here τ_{cm} is the total torque with respect to the center of mass, and $\mathbf{L}_{cm}$ is the angular momentum of the body with respect to the center of mass, that is, an angular momentum of the body with respect to the center of mass in an *inertial* frame moving with the center of mass. $\mathbf{L}_{cm}$ transforms as a three-vector under rotation. Under pure Galilean transformation—no relative rotation of the two frames—the components of $\mathbf{L}_{cm}$ do not change. A particularly simple but important case is $\tau_{cm} = 0$, for which $\mathbf{L}_{cm}$ is constant. It is constant irrespective of how the center of mass is moving! Such a spinning body, with $\tau_{cm} = 0$, is called a *gyroscope.*

What are the dynamics of a gyroscope in special relativity? In the rest frame of the gyroscope, the angular momentum $\mathbf{L}_{cm}$ is characterized by a three-vector with some magnitude and direction, or a four-vector that in this frame has a vanishing time component, that is, $L_\alpha = (0, \mathbf{L}_{cm})$. The existence of such a spacelike four-vector describing the intrinsic angular momentum of a rigid body can be made more explicit by considering the relativistic generalization of angular momentum. Here we take it as a given. The relativistic dynamics of a gyroscope is then easy to describe. Thus, the proper time dependence of $L_\alpha(\tau)$ is such that in the instantaneous inertial rest frame of the gyroscope,

$$\frac{d\mathbf{L}_{cm}}{dx_0} = \frac{d\mathbf{L}_{cm}}{d\tau} = 0. \qquad (3.17)$$

This is assured if

$$\frac{dL_\alpha(\tau)}{d\tau} = SU_\alpha, \qquad (3.18)$$

where S is a scalar and U_α is the instantaneous four-velocity of the gyroscope. This equation sets a four-vector equal to a four-vector. Thus, if correct in one inertial frame, it is correct in any other, and in the gyroscope's inertial frame the space components reduce to Eq. (3.17). What is the scalar S? We will see that it *cannot* generally be zero. Since $L_\alpha \eta_{\alpha\beta} U_\beta = 0$ is true in the gyroscope's inertial frame, it is true in any frame. We have then

$$\frac{dL_\alpha}{d\tau} \eta_{\alpha\beta} U_\beta + L_\alpha \eta_{\alpha\beta} \frac{dU_\beta}{d\tau} = 0.$$

This, with Eq. (3.18) (recall that $U_\alpha \eta_{\alpha\beta} U_\beta = 1$), implies that

$$S = -L_\alpha \eta_{\alpha\beta} \frac{dU_\beta}{d\tau}, \tag{3.19}$$

which, used in Eq. (3.18), leads to

$$\frac{dL_\alpha(\tau)}{d\tau} = -L_\sigma \eta_{\sigma\beta} \frac{dU_\beta}{d\tau} U_\alpha. \tag{3.20}$$

A knowledge of the world line of the gyroscope, $x_i(x_0)$, gives an equation governing how L_α changes, and it does indeed change. If, for example, the gyroscope moves in a circle, L_α will undergo a change with each revolution. Even though in the instantaneous inertial frame of the gyroscope there is no rotation, we see that the "frame" of the gyroscope rotates. That is, if one considers a continuous set of inertial frames tangent to each point on the circle and moving with relative velocity zero to the gyroscope such that nearby inertial frames are oriented without relative rotation (they are "parallel"), there is a rotation of the frame at the completion of the circle relative to the frame at the beginning. (See Exercise 7.) This effect is known as the *Thomas precession*. A gyroscope moving in a circle with a constant speed β precesses through an angle $2\pi(\gamma - 1)$ in a direction opposite to the path rotation. (See Exercise 6.)

3.7 Exercises

1. Show that for two massive particles, with masses m_1 and m_2 and four-momenta $p_\mu^{(1)}$ and $p_\mu^{(2)}$ the sum of their four-momentum, $P_\mu = p_\mu^{(1)} + p_\mu^{(2)}$, is such that $P_\mu n_{\mu\nu} P_\nu \geq (m_1 + m_2)^2$, with equality obtaining only if the velocities of the two particles are equal. (*Hint:* Consider the invariant in the rest frame of one of the particles.)

2. Two particles with momenta (expressed in energy units of Mev) $(3, 1, 1, 0)$ and $(2, 1, -1, 0)$ collide to produce two photons with momenta $(1, 1, 0, 0), (2, 1, 1, \sqrt{2})$, and an unknown particle. (a) What is the momentum of the unknown particle? (b) What is its mass? (c) What is the velocity of the center of momentum frame?

3. *Compton scattering.* A photon of frequency ν scatters off a charged particle of mass m that is at rest, and the scattered photon makes an angle θ with the direction of the incoming photon. Show that

$$\frac{1}{\nu'} = \frac{1}{\nu} + h \frac{1 - \cos(\theta)}{m},$$

where ν' is the frequency of the outgoing photon. Use $p_0 = h\nu$.

4. Show that the total momentum of two photons is timelike unless the photons' directions are parallel.

5. Use the definition of $\mathbf{f}$, Eq. (3.11), to show that

$$\mathbf{f} = \mathbf{f} \cdot \boldsymbol{\beta}\boldsymbol{\beta} + \gamma m\mathbf{a}.$$

Thus, generally, the three-acceleration is not parallel to the three-force. For what cases are they parallel?

6. A gyroscope moves counterclockwise in a circle, in the x_1-x_2 plane, of radius R with a constant speed β. (a) Show that the four-angular momentum L_α of the gyroscope satisfies the equations

$$\frac{dL_0}{d\theta} = -[L_1 \sin\theta + L_2 \cos\theta]\beta\gamma^2$$

$$\frac{dL_1}{d\theta} = [L_1 \sin\theta + L_2 \cos\theta](\beta\gamma)^2 \sin\theta$$

$$\frac{dL_2}{d\theta} = -[L_1 \sin\theta + L_2 \cos\theta](\beta\gamma)^2 \cos\theta.$$

Here θ is the angular position in the circle measured from the x_1 axis.

Now let $L_1 = L_1^0$ and $L_2 = L_0 = 0$ at $\theta = 0$. (b) At $\theta = \Delta\theta$, what is $L_\alpha(\Delta\theta)$ to first order in $\Delta\theta$? (c) What are the components $L'_\alpha(\Delta\theta)$ in the "parallel" frame moving with the gyroscope at $\theta = \Delta\theta$, again to first order in $\Delta\theta$? [Use Eqs. (1.26) and (1.27).] (d) From (c), find through what angle the gyroscope precessed while traveling $\Delta\theta$. Through what angle will the gyroscope precess while traveling one loop?

7. Consider two frames, $F_{(1)}(\beta)$ and $F_{(2)}(\beta + \Delta\beta)$ moving parallel to the frame F with velocities β and $\beta + \Delta\beta$, respectively, with $\beta \perp \Delta\beta$. (a) What are the $F_{(2)}$ coordinates of an event expressed in terms of those of F to first order in $\Delta\beta$? Use Eqs. (1.26) and (1.27). (b) By use of the inverse of Eqs. (1.26) and (1.27), express the $F_{(2)}$ coordinates of an event in terms of those of $F_{(1)}$ and show

$$\mathbf{r}_{(2)} = \mathbf{r}_{(1)} + \frac{\boldsymbol{\phi}}{\beta^2} \times \mathbf{r}_{(1)} - \Delta\boldsymbol{\beta}' x_{(1)0},$$

where $\boldsymbol{\phi} = (\gamma - 1)\Delta\boldsymbol{\beta} \times \boldsymbol{\beta}$ and $\Delta\boldsymbol{\beta}'$ is the velocity of $F_{(2)}$ with respect to $F_{(1)}$ (as viewed from $F_{(1)}$). Thus we see that, though the frames $F_{(2)}$ and $F_{(1)}$ are "parallel" to F, they are rotated with respect to each other. (c) Relate this rotation to the precession of the gyroscope in Exercise 6.

Chapter 4

Relativity of Tensor Fields

4.1 Introduction

So far we have dealt mainly with kinematic and dynamic quantities that arise in the description of the motion of point particles. However, many quantities used in the description of physical phenomena are "fields," which are defined at all positions and for all times, that is, at all events. Examples are the electric field, a three-vector field, and pressure, a three-scalar field. The "three" here refers to the three space dimensions. We will characterize transformation properties of such fields. By differentiating such fields, one forms other fields. As an example, the gradient of a scalar electric potential gives the electric field. We generalize these three-field concepts to four-fields and extend the idea of "gradient" to include its action on four-fields. Here, of course, "four" refers to the four space-time dimensions. We then discuss some particularly important four-fields, namely electromagnetic fields and energy-momentum fields.

4.2 Transformations of Tensors

4.2.1 Three-Tensors

Recall that the rotation of the coordinate system is characterized by a transformation of the coordinates

$$x_i' = \sum_{j=1}^{3} A_{ij} x_j = \frac{\partial x_i'}{\partial x_j} x_j \qquad (4.1)$$

such that

$$\sum_{i=1}^{3} x_i'^2 = \sum_{j=1}^{3} x_j^2. \tag{4.2}$$

We saw that A_{ij} satisfies

$$\delta_{mn} A_{jm} A_{kn} = \delta_{jk}$$
$$\tilde{A}A = I. \tag{4.3}$$

We defined the elements of the rotation group to be any A_{ij} that satisfies Eq. (4.3), A three-vector $\mathbf{a}$ is defined by its transformation,

$$a_i' = \sum_{j=1}^{3} A_{ij} a_j = \frac{\partial x_i'}{\partial x_j} a_j,$$

and, in analogy to a vector, a vector field $\mathbf{a}(\mathbf{r})$, is defined by the transformation

$$a_i'(x') = A_{ij} a_j(x). \tag{4.4}$$

One defines a three-tensor field of rank n as a field $T_{i...k}(x)$, with n indices, with the transformation

$$T_{l...n}'(x') = A_{li}...A_{nk} T_{i...k}(x). \tag{4.5}$$

We see, from Eq. (4.3), that δ_{lm}, the metric, transforms as a three-tensor of rank 2. Viewed as a field, the metric has trivial "event" dependence. An important tensor, sometimes referred to as the Levi-Civita tensor, is the completely antisymmetric tensor of rank 3, ϵ_{ijk}, defined by

$$\epsilon_{ijk} = \begin{cases} 1 & \text{if } ijk \text{ is an even permutation of } 1\,2\,3 \\ -1 & \text{if } ijk \text{ is an odd permutation of } 1\,2\,3 \\ 0 & \text{if any two of } ijk \text{ are equal} \end{cases} \tag{4.6}$$

That the transformed ϵ_{ijk} is completely antisymmetric is easy to see; that $\epsilon_{123} = 1$ is less so. In fact, $\epsilon_{ijk}' = \text{Det } A\, \epsilon_{ijk}$. It follows from Eq. (4.3) that $\text{Det } A = \pm 1$.[1] (See Exercise 1.)

Several operations on tensors result in new tensors. The *outer product* of two tensors, for instance $C_{ijk} = A_{ij} B_k$, form a tensor of a rank that is the sum of the two tensors. *Contraction*—summing over two

---

[1] A "proper" rotation, one that can be effected by rotating about some direction through some finite angle, will satisfy $\text{Det } A = 1$. For an "improper" rotation, such as $x_i' = -x_i$, $\text{Det } A = -1$.

repeated indices—results in a tensor of rank 2 less. An example of such an operation is the cross-product of two vectors **A** and **B**, resulting in a vector **C** given by

$$C_i = \epsilon_{ijk} A_j B_k.$$

An important operation performed on tensor fields is *differentiation*. We ask then how $\partial a_i(x)/\partial x_j$ transforms; that is, what is the relation of $\partial a_i'(x')/\partial x_j'$ to $\partial a_i(x)/\partial x_j$? This relation is determined by the transformation properties of $\mathbf{a}(\mathbf{r})$ and of x_i. We have

$$\frac{\partial a_i'(x')}{\partial x_k'} = A_{ij} \frac{\partial a_j(x)}{\partial x_k'} = A_{ij} \frac{\partial x_m}{\partial x_k'} \frac{\partial a_j(x)}{\partial x_m}, \qquad (4.7)$$

where use has been made of the chain rule

$$\frac{\partial}{\partial x_k'} = \frac{\partial x_m}{\partial x_k'} \frac{\partial}{\partial x_m}.$$

But

$$A_{km} x_k' = A_{km} A_{kj} x_j = x_m,$$

and thus

$$\frac{\partial x_m}{\partial x_k'} = A_{km},$$

which merely restates that the inverse of **A** is its transpose, $\tilde{\mathbf{A}}$. Using this result in Eq. (4.7), we obtain

$$\frac{\partial a_i'(x')}{\partial x_k'} = A_{ij} A_{km} \frac{\partial a_j(x)}{\partial x_m}. \qquad (4.8)$$

We see that $\partial a_j(x)/\partial x_m$ transforms as a tensor of rank 2. The operation $\partial_m \equiv \partial/\partial x_m$ raises the rank of a tensor field by one, adding a vector index m. A familiar example of this, the gradient of a scalar field, results in a vector field.

This result, that operation by ∂_m on a tensor field produces a new tensor field of one higher rank, depends on the property of rotations $A_{ij} A_{ik} = \delta_{jk}$, which is not shared by Lorentz transformations.

4.2.2   Four-Tensors

Before we discuss the transformation properties of four-tensors and the effect on transformation properties by action of partial derivatives, we introduce an important change in notation that agrees with traditional usage. Rather than indicating the space-time coordinates of an event by x_μ, we will use x^μ. That is, we will use superscripts rather than subscripts. Thus, a Lorentz transformation will be written as

$$x'^\mu = \Lambda^{\mu'}_\beta x^\beta, \qquad (4.9)$$

where $\Lambda^{\mu'}_\beta$ satisfies[2]

$$\Lambda^{\mu'}_\beta \, n_{\mu\delta} \, \Lambda^{\delta'}_\rho = n_{\beta\rho}$$
$$\tilde{\Lambda} \mathbf{n} \Lambda = \mathbf{n}. \qquad (4.10)$$

Here $n_{\beta\rho}$ is the metric introduced earlier, which we write with subindices. Note that the repeated indices are one subscript and one superscript. We will maintain this convention. Also, we have introduced a somewhat mixed notation; in writing primed event coordinates, the "x"s are primed, not the index; the index merely takes on the values $0, 1, 2, 3$. However, in writing the coefficients Λ we prime the index to indicate the index of the primed coordinates. In the transformation equation for coordinates, Eq. (4.9), the sum is over the lower index of Λ, an unprimed coordinate. The coefficients $\Lambda^{\mu'}_\beta$ implement the Lorentz transformation from the unprimed coordinates to the primed coordinates. The coefficients $\Lambda^\mu_{\beta'}$ implement the transformation from the primed coordinates to the unprimed coordinates. The distinction is easily recognized by the notation.

The transformation of a contravariant (four) vector field with components $V^\mu(x)$ is given by

$$V'^\mu(x') = \Lambda^{\mu'}_\beta V^\beta(x) = \frac{\partial x'^\mu}{\partial x^\beta} V^\beta(x), \qquad (4.11)$$

which gives the components of the vector in the primed frame at the same event point. The designation "contravariant vector" is sometimes used. We will see there is reason to introduce a different type of "vector" field called a *covector*—sometimes a covariant vector field.

How does $\partial V^\beta(x)/\partial x^\mu$ transform? That is, what is the relation between $\partial V'^\beta(x')/\partial x'^\mu$ and $\partial V^\beta(x)/\partial x^\mu$, something surely determined by the transformations of V and x. We have

[2] Similar to rotations, not all $\Lambda's$ satisfying Eq. (4.10) can be built up by a sequence of small Lorentz boosts and "proper" rotations and thus are not all smoothly connected to the identity. Again, $\text{Det}\,\Lambda = \pm 1$. In addition, from Eq. (4.10) it is easy to see that $|\Lambda^0_0| \geq 1$. To be smoothly connected to the identity, then, $\text{Det}\,\Lambda = +1$ and Λ^0_0 is positive. The Lorentz group with such restrictions is referred to as the *restricted Lorentz group*.

$$\frac{\partial V'^{\beta}(x')}{\partial x'^{\mu}} = \Lambda_{\rho}^{\beta'}\frac{\partial V^{\rho}}{\partial x'^{\mu}} = \Lambda_{\rho}^{\beta'}\frac{\partial x^{\sigma}}{\partial x'^{\mu}}\frac{\partial V^{\rho}}{\partial x^{\sigma}} = \Lambda_{\rho}^{\beta'}\Lambda_{\mu'}^{\sigma}\frac{\partial V^{\rho}}{\partial x^{\sigma}}. \qquad (4.12)$$

We have set $\partial x^{\sigma}/\partial x'^{\mu} = \Lambda_{\mu'}^{\sigma}$ where

$$x^{\sigma} = \Lambda_{\mu'}^{\sigma} x'^{\mu},$$

the "inverse" of the original Lorentz transformation. Using the chain rule of differentiation, we find

$$\frac{\partial x'^{\mu}}{\partial x'^{\rho}} = \delta_{\rho}^{\mu} = \frac{\partial x'^{\mu}}{\partial x^{\beta}}\frac{\partial x^{\beta}}{\partial x'^{\rho}} = \Lambda_{\beta}^{\mu'}\Lambda_{\rho'}^{\beta}. \qquad (4.13)$$

Here $\Lambda_{\rho'}^{\beta}$ is the "matrix" inverse of $\Lambda_{\beta}^{\mu'}$. We can ask how $\Lambda_{\beta}^{\mu'}$ can be expressed in terms of $\Lambda_{\rho'}^{\beta}$. First, define $n^{\rho\sigma}$ so that $n^{\rho\sigma}n_{\sigma\beta} = \delta_{\beta}^{\rho}$ ($n^{\rho\sigma} = \text{Diag}\{1, -1, -1, -1\}$). Then consider the Lorentz transformation

$$x'^{\mu} = \Lambda_{\beta}^{\mu'} x^{\beta}. \qquad (4.14)$$

By use of Eq. (4.10), we can write

$$n^{\rho\sigma}\Lambda_{\sigma}^{\alpha'} n_{\alpha\mu}\Lambda_{\beta}^{\mu'} = \delta_{\beta}^{\rho}$$

and thus obtain

$$x^{\rho} = n^{\rho\sigma}\Lambda_{\sigma}^{\alpha'} n_{\alpha\mu}\Lambda_{\beta}^{\mu'} x^{\beta} = n^{\rho\sigma}\Lambda_{\sigma}^{\alpha'} n_{\alpha\mu}x'^{\mu}.$$

From this it follows that

$$\Lambda_{\mu'}^{\rho} = \frac{\partial x^{\rho}}{\partial x'^{\mu}} = n^{\rho\sigma}\Lambda_{\sigma}^{\alpha'} n_{\alpha\mu}. \qquad (4.15)$$

With

$$A_{\sigma}^{\rho}(x) \equiv \frac{\partial V^{\rho}(x)}{\partial x^{\sigma}},$$

the upper index of A transforms as dx^{ρ} (a contravariant index), whereas the lower index transforms differently, as $\frac{\partial}{\partial x^{\sigma}} \equiv \partial_{\sigma}$ (covariant index). Thus, we have

$$A_{\alpha}'^{\beta}(x') = \Lambda_{\rho}^{\beta'}\Lambda_{\alpha'}^{\sigma}A_{\sigma}^{\rho}(x). \qquad (4.16)$$

The definition of a mixed tensor of covariant rank n and contravariant rank m is clear. A has covariant rank 1 and contravariant rank 1. A particular example of a covariant vector (a covariant tensor of rank 1) is the "gradient" of a scalar field $\theta(x)$,

$$\partial'_\mu \, \theta'(x') = \Lambda^\beta_{\mu'} \partial_\beta \theta(x).$$

Note that, from Eq. (4.10), the metric $n_{\mu\delta}$ transforms as tensor of covariant rank 2; $\Lambda^{\mu'}_\beta$ effects the Lorentz transformation from x to x'. Furthermore, $n^{\beta\rho}$ transforms as a tensor of contravariant rank 2.

As for three-tensors, an important tensor, again sometimes called the Levi-Civita tensor, is the completely antisymmetric tensor of contravariant rank 4 $\epsilon^{\alpha\beta\rho\sigma}$ given by

$$\epsilon^{\alpha\beta\rho\sigma} = \begin{cases} 1 & \text{if } \alpha\beta\rho\sigma \text{ is an even permutation of } 0\,1\,2\,3 \\ -1 & \text{if } \alpha\beta\rho\sigma \text{ is an odd permutation of } 0\,1\,2\,3 \\ 0 & \text{if any two of } \alpha\beta\rho\sigma \text{ are equal} \end{cases} \qquad (4.17)$$

And as for the completely antisymmetric three-tensor, there is a caveat to be added; actually,

$$\epsilon'^{\alpha\beta\rho\sigma} = \text{Det } \Lambda \, \epsilon^{\alpha\beta\rho\sigma}.$$

Just as in three-space, the *outer product* of two tensors, for instance $A_{\rho\beta} B^\sigma$, forms a tensor of ranks which are the sums of the ranks of the two tensors.

Again, *contraction* can be used to form a new tensor—here, however, the summing must be over an upper and lower index. One can see that such an operation produces a tensor of one lower contravariant rank and one lower covariant rank. For example, consider

$$B'^\mu_{\mu\beta}(x') = \Lambda^{\mu'}_\sigma \Lambda^\rho_{\mu'} \Lambda^\alpha_{\beta'} B^\sigma_{\rho\alpha}(x) = \delta^\rho_\sigma \Lambda^\alpha_{\beta'} B^\sigma_{\rho\alpha}(x) = \Lambda^\alpha_{\beta'} B^\sigma_{\sigma\alpha}(x).$$

Eq. (4.13) has been used. A particularly important contraction operation is the raising (lowering) of indices by contraction with $n^{\alpha\rho}$ (with $n_{\alpha\rho}$). As an example, given B_ρ we define $B^\alpha = n^{\alpha\rho} B_\rho$. Note $dx_\mu \, dx^\mu = dx^\alpha \, n_{\alpha\mu} \, dx^\mu$ is the invariant interval.

4.3   Relativity of Maxwell's Equations

The Maxwell equations are invariant in form under Lorentz transformations. This had been discovered by Lorentz, but he did not interpret their physical meaning in a way that gave them relevance outside of electromagnetic theory. Thus, these equations need no modification to make them compatible with Einsteinian relativity. We now examine the Maxwell equations themselves to deduce the field transformation properties.

We begin with the continuity equation,

$$\frac{\partial \rho}{\partial t} + \nabla \cdot \boldsymbol{J} = 0. \tag{4.18}$$

Here ρ is a electric charge density and $\boldsymbol{J}$ is the electric current density. It is natural to postulate that $(\rho, \boldsymbol{J}/c)$ form a contravariant vector field J^{α}, for the continuity equation can then be written as

$$\partial_{\alpha} J^{\alpha} = 0. \tag{4.19}$$

Consider now the inhomogeneous equations,

$$\nabla \cdot \mathbf{E} = 4\pi\rho \tag{4.20}$$

$$\nabla \times \mathbf{B} - \frac{1}{c}\frac{\partial \mathbf{E}}{\partial t} = \frac{4\pi}{c}\,\mathbf{J}. \tag{4.21}$$

We have postulated that the right-hand sides of these equations constitute the time and space components of a four-vector. If these equations are to be true in all frames, the left-hand sides must form a four-vector too. But on the left-hand sides, the action by ∂_{μ}, which raises the covariant index by one unless contracted, is involved. Thus, these equations must be of the form

$$\partial_{\alpha} F^{\alpha\beta} = 4\pi J^{\beta}. \tag{4.22}$$

This $F^{\alpha\beta}$, called the *field-strength tensor*, is a contravariant tensor of rank 2 and linear in the **B** and **E** fields. Since there are but six components to these fields, one expects $F^{\alpha\beta}$ is antisymmetric in the two indices, thus having six independent components. The time component of Eq. (4.22) becomes Eq. (4.20) with the identification $F^{i0} = E^{i}$. Eq. (4.21) implies that the components B^{i} must be in the space-space parts F^{ij}. It is easy to check that

$$F^{\alpha\beta} = \begin{bmatrix} 0 & -E^{1} & -E^{2} & -E^{3} \\ E^{1} & 0 & -B^{3} & B^{2} \\ E^{2} & B^{3} & 0 & -B^{1} \\ E^{3} & -B^{2} & B^{1} & 0 \end{bmatrix} \tag{4.23}$$

satisfies Eq. (4.22). The components of the **E** and **B** fields are written with upper indices to agree with the convention now being used of writing the components of the position **r** and time x^0 with upper indices.

What about the homogeneous Maxwell's equations,

$$\nabla \cdot \mathbf{B} = 0 \tag{4.24}$$

$$\nabla \times \mathbf{E} + \frac{1}{c}\frac{\partial \mathbf{B}}{\partial t} = 0? \tag{4.25}$$

If $F^{\alpha\beta}$ transforms as a contravariant tensor of rank 2, will these equations be valid in all frames if valid in one? That is, can they be put into a (Lorentz) invariant form? If one defines the antisymmetric tensor of contravariant rank 2 by

$$\mathcal{F}^{\alpha\beta} = \frac{1}{2}\,\varepsilon^{\alpha\beta\rho\sigma}\,F_{\rho\sigma} = \begin{bmatrix} 0 & -B^1 & -B^2 & -B^3 \\ B^1 & 0 & E^3 & -E^2 \\ B^2 & -E^3 & 0 & B^1 \\ B^3 & E^2 & -E^1 & 0 \end{bmatrix} \tag{4.26}$$

(said to be the "dual" tensor of $F^{\rho\sigma}$), one can see that the homogeneous Maxwell equations can be written

$$\partial_\alpha \mathcal{F}^{\alpha\beta} = 0, \tag{4.27}$$

which can also be expressed as

$$\partial_\alpha F_{\beta\gamma} + \partial_\gamma F_{\alpha\beta} + \partial_\beta F_{\gamma\alpha} = 0. \tag{4.28}$$

The continuity equation and Maxwell's equations can be written in invariant form. Can the Lorentz force law that gives the interaction of the electromagnetic field, a tensor of rank 2, and the charge density and three-current density, which taken together form a four-vector, be expressed invariantly? Considering that Newton's force law cannot be put in invariant form, one may expect some difficulty. First, consider the relativistic form of Newton's force law for a charged particle with charge q in an electromagnetic field. In the rest frame of the charged particle—see Eq. (3.15)—we have

$$F^\alpha = (0, q\mathbf{E}) = \left(0, \frac{d\mathbf{p}}{dx^0}\right). \tag{4.29}$$

Any four-vector that reduces to this in the rest frame must be the four-force for a moving charged particle:

$$F^\alpha = qn_{\rho\beta} U^\rho F^{\alpha\beta} = \gamma(x)(q\boldsymbol{\beta} \cdot \mathbf{E}, q\mathbf{E} + e\boldsymbol{\beta} \times \mathbf{B}). \qquad (4.30)$$

We deduce from this—recall Eq. (3.13)—that the three-force is given by

$$\mathbf{f} = q\mathbf{E} + q\boldsymbol{\beta} \times \mathbf{B}. \qquad (4.31)$$

This Lorentz force is determined by the transformation properties of the electromagnetic field and the form of the force in the rest frame of the particle. By writing the charge density and current density for point particles, one can deduce the Lorentz force per unit volume:

$$\mathbf{f}(x) = \rho\mathbf{E}(x) + \mathbf{J}(x) \times \mathbf{B}(x). \qquad (4.32)$$

This is the space part of a four-vector field given by

$$f^\beta(x) = F^{\alpha\beta}(x) J_\beta. \qquad (4.33)$$

4.4 Dynamics of a Charged Spinning Particle

The study of the dynamics of a charged spinning particle is of special interest since many charged "elementary" particles have a spin (intrinsic) angular momentum. Even though the spin is a quantum mechanical property of the particle, some insight into the dynamics of spin can be obtained by treating the spin as a classical angular momentum of fixed magnitude—a gyroscope. (See Sec. 3.6.) In addition, the power of invariance arguments is well illustrated by their study. A rotating spherical charged particle has associated with it a magnetic moment μ that points parallel/antiparallel to the "intrinsic" angular momentum S. When such a particle is at rest in the presence of a magnetic field, a torque is exerted on the particle so that

$$\frac{d\mathbf{S}}{dx^0} = \boldsymbol{\mu} \times \mathbf{B} = \frac{gq}{2m} \mathbf{S} \times \mathbf{B}. \qquad (4.34)$$

Here q is the charge and m is mass of the "particle." The constant g is called the gyromagnetic ratio. For a spinning body whose charge density to mass density ratio is constant, $q = 1$.

What is the dynamical equation for the spin of a charged particle in special relativity? That is, what form-invariant equation reduces to

Eq. (4.34) in the rest frame of the particle? As in the case of the angular momentum vector of the gyroscope, we take it as given that the spin angular momentum S^α is a four-vector that has a vanishing time-component in the rest frame of the particle and thus satisfies

$$S^\alpha U_\alpha = 0. \tag{4.35}$$

For such a four-vector, in the rest frame of the particle, we have

$$\frac{gq}{2m} S_\beta F^{\alpha\beta} = \frac{gq}{2m} (\mathbf{E} \cdot \mathbf{S}, \mathbf{S} \times \mathbf{B}). \tag{4.36}$$

Consider the equation

$$\frac{dS^\alpha}{d\tau} = C U^\alpha + \frac{gq}{2m} S_\beta F^{\alpha\beta}, \tag{4.37}$$

with C a scalar. (Recall Eq. (3.18) for the gyroscope.) In the rest frame of the particle, the space components of Eq. (4.37) reduce to Eq. (4.34).

By use of Eqs. (4.35) and (4.37) we find

$$C = - \left[S^\rho \frac{dU_\rho}{d\tau} + \frac{gq}{2m} U_\rho F^{\rho\beta} S_\beta \right]. \tag{4.38}$$

With this expression for C, Eq. (4.37) becomes

$$\frac{dS^\alpha}{d\tau} = - \left[S^\rho \frac{dU_\rho}{d\tau} + \frac{gq}{2m} U_\rho F^{\rho\beta} S_\beta \right] U^\alpha + \frac{gq}{2m} S_\beta F^{\alpha\beta}, \tag{4.39}$$

If the electromagnetic field tensor vanishes, Eq. (4.39) becomes the gyroscope equation, Eq. (3.20). If the electromagnetic field four-force, Eq. (4.30), alone acts on the charged particle, Eq. (4.39) can be written as

$$\frac{dS^\alpha}{d\tau} = - \frac{q}{m} \left[\left(1 - \frac{g}{2} \right) S_\rho F^{\rho\beta} U_\beta \right] U^\alpha + \frac{gq}{2m} S_\beta F^{\alpha\beta}. \tag{4.40}$$

In this case, the Thomas precession term $(S^\rho dU_\rho / d\tau) U^\alpha$ is partially canceled. If the particle has a gyromagnetic ratio of two, the Thomas precession term is completely canceled, resulting in the equation

$$\frac{dS^\alpha}{d\tau} = \frac{q}{m} S_\beta F^{\alpha\beta}. \tag{4.41}$$

In 1926, Uhlenbeck and Goudsmit introduced the idea of spin of an elementary particle and argued that if the electron had a gyromagnetic ratio of two, various effects of line spectra could be explained. Finally, the quantum relativistic equation of Dirac results in a gyromagnetic ratio of two for the particle it describes, thus giving a relativistic basis for the value postulated by Uhlenbeck and Goudsmit to explain line spectra.

4.5   Local Conservation and Gauss's Theorem

The continuity equation, Eq. (4.18), implies the local conservation of charge, that is, that the rate of change charge in a three-volume V_3 equals the rate at which the charge current **J** brings charge into the volume. This is seen by an application of the three-dimensional Gauss's theorem:

$$\int_{V_3} \nabla \cdot \mathbf{A} dv = \int_{\partial V_3} \mathbf{A} \cdot d\mathbf{S}.$$

Here ∂V indicates the closed two-dimensional surface (boundary) of the volume, and $d\mathbf{S}$ is the outward directed element of the surface. Thus, Eq. (4.19) yields

$$\frac{d}{dt} \int_{V_3} \rho dv = - \int_{\partial V_3} \mathbf{J} \cdot d\mathbf{S} = - \int_{\partial V_3} \mathbf{J} \cdot \mathbf{n} d\mathbf{S}.$$

The change ΔQ in the charge in the volume between t_0 and t_1 is given by

$$\Delta Q = \int_{V_3} \rho(t_1, \mathbf{r}) dv - \int_{V_3} \rho(t_0, \mathbf{r}) dv = - \int_{t_0}^{t_1} \int_{\partial V_3} \mathbf{J} \cdot \mathbf{n} d\mathbf{S} dt. \qquad (4.42)$$

If V_3 is all of three-space, and if $\mathbf{J} \neq 0$ only in a compact region of space, then $\Delta Q = 0$. Local conservation of charge, Eq. (4.18), implies global conservation of charge.

Let us now apply the four-dimensional Gauss's theorem to the invariant form of the continuity equation, Eq. (4.19):

$$\int_{V_4} \partial_\alpha J^\alpha dx^0 dx^1 dx^2 dx^3 = \int_{\partial V_4} J^\alpha dS_\alpha$$

$$= \int_{\partial V_4} (J^0 dS_0 + J^1 dS_1 + J^2 dS_2 + J^3 dS_3) = 0. \qquad (4.43)$$

But what is the covariant vector dS_α? Consider a four-volume $V_3 \Delta x^0$, with $V_3 = \Delta x^1 \Delta x^2 \Delta x^3$ depicted in Figure 4.1. The three-dimensional Gauss's theorem applied to Eq. (4.19) for this four-volume gives

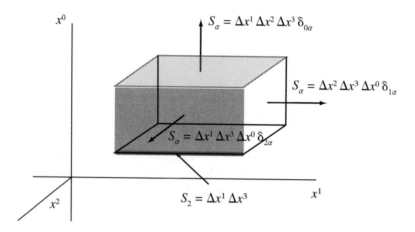

Figure 4.1. Four-volume V_4.

$$\int_{V_4} \partial_\alpha J^\alpha \, dx^0 \, dx^1 \, dx^2 \, dx^3 = \int_{x_1^0}^{x_2^0} \int_{\partial V_3} \mathbf{J} \cdot d\mathbf{S} \, dx^0 + \int_{V_3} J^0(x_2^0, \mathbf{r}) \, dx^1 \, dx^2 \, dx^3$$

$$- \int_{V_3} J^0(x_1^0, r) \, dx^1 \, dx^2 \, dx^3 = 0. \qquad (4.44)$$

By use of this and Eq. (4.43), for this four-volume, we can identify the particular surface element dS_α associated with the various parts of the three-surface boundary:

$$dS_\beta = \begin{cases} (dx^1 \, dx^2 \, dx^3, 0, 0, 0) & \text{for the surface at } x_2^0 \\ -(dx^1 \, dx^2 \, dx^3, 0, 0, 0) & \text{for the surface at } x_1^0 \\ (0, Sdx^0) & \text{for the four ``vertical'' sides of the surface.} \end{cases} \qquad (4.45)$$

We can write these surface elements in a manner that illustrates their covariant vector transformation properties. Consider, for example, the surface element $dS_\beta = (dx^1 \, dx^2 \, dx^3, 0, 0, 0)$. It is determined by three four-vectors, namely $da^\mu = (0, dx^1, 0, 0)$, $db^\mu = (0, 0, dx^2, 0)$, and $dc^\mu = (0, 0, 0, dx^3)$. The four-vector dS_β must result from contraction of these three four-vectors with a tensor of covariant rank 4. Thus, we have

$$dS_\beta = \epsilon_{\beta\alpha\rho\sigma} \, da^\alpha \, db^\rho \, dc^\sigma. \qquad (4.46)$$

Here $\epsilon_{\beta\alpha\rho\sigma}$ is the completely antisymmetric tensor of covariant rank-4. A general three-dimensional surface element dS_β characterized by three (ordered) four-vectors dx_1^α, dx_2^ρ, and dx_3^σ is given by

$$dS_\beta = \epsilon_{\beta a \rho \sigma} \, dx_1^a \, dx_2^\rho \, dx_3^\sigma. \tag{4.47}$$

As another example, a general surface element of the "vertical" sides of our three-surface is characterized by the three four-vectors, $da^\mu = (dx^0, 0, 0, 0)$, $db^\mu = (0, d\mathbf{B})$, and $dc^\mu = (0, d\mathbf{C})$. This results in

$$dS_\beta = \epsilon_{\beta a \rho \sigma} \, da^a \, db^\rho \, dc^\sigma = (0, (d\mathbf{B} \times d\mathbf{C}) \, dx^0) = (0, d\mathbf{S} dx^0). \tag{4.48}$$

We can, of course, apply the four-dimensional Gauss's theorem to the four-divergence of any four-tensor field, $T^{a\beta\ldots\sigma}(x)$. Thus,

$$\int_{v_4} \partial_\sigma T^{a\beta\ldots\sigma}(x) = \int_{\partial V_4} T^{a\beta\ldots\sigma}(x) \, dS_\sigma. \tag{4.49}$$

If $\partial_\sigma T^{a\beta\ldots\sigma}(x) = 0$, this gives

$$\int_{\partial V_4} T^{a\beta\ldots\sigma}(x) \, dS_\sigma = 0,$$

which, if V_4 is our volume of Figure 4.1, gives

$$\int_{\partial V_4} T^{a\beta\ldots\sigma}(x) \, dS_\sigma = \int_{x_1^0}^{x_2^0} \int_{\partial V_3} T^{a\beta\ldots i}(x) \, dS_i \, dx^0 + \int_{V_3} T^{a\beta\ldots 0}(x_2^0, \mathbf{r}) \, dx^1 \, dx^2 \, dx^3$$

$$- \int_{V_3} T^{a\beta\ldots 0}(x_1^0, \mathbf{r}) \, dx^1 \, dx^2 \, dx^3 = 0 \tag{4.50}$$

As in the case of the conservation of charge, if V_3 is all of three-space, and if $T^{a\beta\ldots i}(x) \neq 0$ only in a compact region of space, then

$$\int T^{a\beta\ldots 0}(x_2^0, \mathbf{r}) \, dx^1 \, dx^2 \, dx^3 = \int T^{a\beta\ldots 0}(x_1^0, \mathbf{r}) \, dx^1 \, dx^2 \, dx^3. \tag{4.51}$$

$\int T^{a\beta\ldots 0}(x^0, \mathbf{r}) \, dx^1 \, dx^2 \, dx^3$ is conserved.

4.6   Energy-Momentum Tensor

One would expect that for physical phenomena described by fields such as the electromagnetic fields, an energy and momentum should be associated with the fields themselves. Similarly, we should expect that any continuous distribution of matter, for example a gas, should have an energy and momentum associated with it. Further, this field description should permit us to associate an energy and momentum density with a particular point in space at a particular time, and not just a total global energy and momentum. From the required relativistic invariance we discuss what form this association must take.

Consider, then, the momentum dp^i and energy dp^0 that are contained in an infinitesimal volume element $dx^1 dx^2 dx^3$. Here dp^i and dp^0 are the space and time components of a single contravariant four-vector, which we write as dp^α. Clearly, dp^α is proportional to $dx^1 dx^2 dx^3$:

$$dp^\alpha = T^{\alpha\cdots}(x)\, dx^1 dx^2 dx^3 = T^{\alpha\cdots}(x)\, dS_0.$$

Here we have written the three-surface element $dx^1 dx^2 dx^3$ as dS_0, that is, as the time component of a covariant three-vector, as we learned it is in Section 4.5. The requirement that dp^α transform as a vector in turn implies that $T^{\alpha\cdots}(x)$ is a contravariant tensor of rank 2 and thus

$$dp^\alpha = T^{\alpha\beta}(x)\, dS_\beta = T^{\alpha 0}(x)\, dx^1 dx^2 dx^3.$$

We see that $T^{\alpha 0}(x)$ is the α component of the energy-momentum density at an event point x. From our discussion in Section 4.5, we know that if $T^{\alpha\beta}$ satisfies the local conservation equation,

$$\partial_\beta T^{\alpha\beta} = 0, \qquad (4.52)$$

then the total four-momentum,

$$p^\alpha(x^0) = \int T^{\alpha\beta}(x^0, x^1, x^2, x^3)\, dS_\beta = \int T^{\alpha 0}(x^0, x^1, x^2, x^3)\, dx^1 dx^2 dx^3 \qquad (4.53)$$

is conserved, that is it does not depend on x^0. All noninteracting or self-interacting fields should satisfy Eq. (4.52), the "local" conservation law.

We know the physical meaning of $T^{\alpha 0}(x)$. What is the physical meaning of the remaining components, $T^{\alpha j}(x)$? In order to answer this question, let us consider an infinitesimal four-volume defined by the four-vectors $(dx^0, 0, 0, 0)$, $(0, dx^1, 0, 0)$, $(0, 0, dx^2, 0)$, and $(0, 0, 0, dx^3)$. This volume has eight cubic three-surfaces each characterized by a surface element four-vector, Eq. (4.46). There are two timelike surfaces with four-vectors $(dx^1 dx^2 dx^3, 0, 0, 0)$ and its negative, and six spacelike surfaces with four-vectors $(0, dx^2 dx^3 dx^0, 0, 0)$, $(0, 0, dx^3 dx^0 dx^1, 0)$, $(0, 0, 0, dx^0 dx^1 dx^2)$, and their negatives. Now integrating Eq. (4.52) over the infinitesimal volume and applying the four-dimensional Gauss's theorem, one obtains

$$[T^{\alpha 0}(x^0 + dx^0) - T^{\alpha 0}(x^0)]\, dx^1 dx^2 dx^3 = [T^{\alpha 1}(x^1) - T^{\alpha 1}(x^1 + dx^1)]\, dx^2 dx^3 dx^0$$
$$+ [T^{\alpha 2}(x^2) - T^{\alpha 2}(x^2 + dx^2)]\, dx^3 dx^0 dx^1$$
$$+ [T^{\alpha 3}(x^3) - T^{\alpha 3}(x^1 + dx^3)]\, dx^0 dx^1 dx^2.$$

$$(4.54)$$

From this we see that in addition to T^{00} being the energy density and T^{i0} the momentum density, T^{0i} is the energy flux across a two-surface of constant x^i, and T^{ij} is the momentum flux of component i across a two-surface of constant x^j. This can be due to fields carrying the momentum across and/or stress "forces" being exerted across the surface. Since most readers would be more familiar with applying Gauss's theorem in three dimensions than in four dimensions, the interpretation of momentum flux is perhaps more easily seen by integrating Eq. (4.52) over some *three*-volume and applying Gauss's theorem to the space divergence term:

$$\frac{d}{dx^0}\int_{V_3} T^{\alpha 0}\, dV = -\int_{V_3}\partial_i T^{\alpha i}\, dV = -\int_{\partial V_3} T^{\alpha i}\, dS_i. \tag{4.55}$$

4.6.1   Energy-Momentum Tensor of Dust

Let us first discuss a rather trivial example of an energy-momentum tensor, that of "dust," by which we mean a collection of particles that at each point move together (not randomly) with some velocity. In the frame for which the particles are at rest (for a particular event point) $T_D^{00} = \rho_0$, and, in this frame, the three-momentum density is zero, that is, $T_D^{i0} = 0$. Similarly, since there is no energy flux in any direction, $T_D^{0i} = 0$. And there is no momentum flux (no stresses), since the dust is noninteracting, that is, $T_D^{ij} = 0$. Here ρ_0 is the mass density in the rest frame of the dust and thus, by definition, is a scalar field. Note that $\rho_0 = m_0 n$ where m_0 is the mass of the dust particles, and n is the number density of dust particles. Then, with U^α the four-velocity of the dust in any other frame, we have

$$T_D^{\alpha\beta} = \rho_0(x) U^\alpha(x) U^\beta(x), \tag{4.56}$$

since this is tensor of rank 2 and reduces to the correct form in the rest frame of the dust and is form invariant under Lorentz transformations. From our characterization of dust, we have

$$\partial_\beta T_D^{\alpha\beta} = 0. \tag{4.57}$$

The motion of the dust is governed by the (local) conservation of energy and momentum. That is the conservation of particles of dust, and zero acceleration of elements of the dust fluid is implied by Eq. (4.57). This is easily seen in the instantaneous rest frame of the dust at a particular event point. In such a frame, Eq. (4.57) becomes

$$\partial_\beta(\rho_0(x)U^\alpha(x)U^\beta(x)) = \frac{\partial\rho_0}{\partial x^0}\delta^{\alpha 0} + \rho_0\frac{\partial U^\alpha}{\partial x^0} + \rho_0\frac{\partial U^\beta}{\partial x^\beta}\delta^{\alpha 0}$$

$$= (\frac{\partial\rho_0}{\partial x^0} + \rho_0\nabla\cdot\boldsymbol{\beta})\delta^{\alpha 0} + \rho_0\frac{\partial\beta^{(j)}}{\partial x^0}\delta^{\alpha(j)} \qquad (4.58)$$

$$= 0.$$

In this equation (j) indicates that the j index is not summed. The time component of Eq. (4.58),

$$\frac{\partial\rho_0}{\partial x^0} + \rho_0(x)\nabla\cdot\boldsymbol{\beta} = 0, \qquad (4.59)$$

implies the conservation of mass or the number of particles. The space components of Eq. (4.58),

$$\rho_0(x)\frac{\partial\beta^j}{\partial x^0} = 0, \qquad (4.60)$$

shows the dust has zero acceleration.

4.6.2   Energy-Momentum Tensor of a Perfect Fluid

We will not consider in detail the energy-momentum tensor of a general fluid, but we find it useful to discuss some general properties in order to compare them with the properties particular to a perfect fluid. For a general fluid, then, consider the frame in which the material of the fluid is at rest at a particular event point. At this point, in this particular frame, again $T^{00} = \rho_0$. However, though the energy flux due to the flow of the fluid is zero, heat flow can contribute to energy flux so that, generally, T^{0i} is not zero. Similarly, though the momentum density of the material of the fluid is zero, the heat flow can contribute a momentum density with the result that, generally, T^{i0} is not zero. A perfect fluid can be characterized as one such that at each event point in a frame moving with the fluid the nearby region is seen to be isotropic—all directions are equivalent. In such a frame, $T_F^{j0} = T_F^{0j} = 0$, $\;j = 1, 2, 3$. If T_F^{0i} were not zero, then the direction of energy flux would be special, and if T_F^{j0} were not zero, then the direction of the momentum density would be special. These results imply the there is no heat flow—for a perfect fluid the changes occurring in the fluid are *adiabatic*. Since it is not assumed that the fluid is noninteracting, we cannot conclude that $T_F^{ij} = 0$. However, since all directions are assumed equivalent, momentum flux and stresses must be radial. Thus $T_F^{ij} = p\delta_{ij}$, where p is the *pressure* at the event point. The shear forces are zero. Thus, in this frame $T_F^{\alpha\beta}$ reduces to

$$T_F^{ij} = p\delta^{ij}$$
$$T_F^{i0} = T^{0i} = 0$$
$$T_F^{00} = \rho. \tag{4.61}$$

Here ρ is the *proper relativistic energy density*, the energy density in the rest frame of the fluid.

We can write $T_F^{\alpha\beta}$ in invariant form, which at a given event point reduces to Eq.(4.61) in the rest frame of the fluid. We have at our disposal the fluid's four-velocity $U^\alpha(x)$, the scalar fields $\rho(x)$ and $p(x)$, and the metric $n^{\alpha\beta}$. Thus,

$$T_F^{\alpha\beta} = -p(x)n^{\alpha\beta} + (\rho(x) + p(x))U^\alpha U^\beta, \tag{4.62}$$

which is easily seen to reduce to Eq. (4.61) in the rest frame of the fluid.

As for dust, the motion of the fluid is (partially) governed by the local conservation of energy momentum, Eq. (4.52). To see the physics of this local conservation equation, we evaluate this equation, using the energy-momentum tensor given by Eq. (4.62), in the frame moving with the fluid at the particular event, with the result

$$\partial_\beta T_F^{\alpha\beta} = -\frac{\partial p(x)}{\partial x^\beta} n^{\alpha\beta} + \frac{\partial(\rho(x) + p(x))}{\partial x^0}\delta^{\alpha 0}$$

$$+ (\rho(x) + p(x))\frac{\partial U^\alpha}{\partial x^0} + (\rho(x) + p(x))\frac{\partial U^\beta}{\partial x^\beta}\delta^{\alpha 0}$$

$$= (\frac{\partial\rho}{\partial x^0} + (\rho + p)\nabla\cdot\boldsymbol{\beta})\delta^{\alpha 0} + \left[(\rho + p)\frac{\partial\beta^j}{\partial x^0} + \frac{\partial p}{\partial x^j}\right]\delta^{\alpha j}$$

$$= 0. \tag{4.63}$$

The time component of Eq. (4.63) can be written as

$$\frac{\partial\rho}{\partial x^0} + (\rho + p)\nabla\cdot\boldsymbol{\beta} = \frac{\partial\rho}{\partial x^0} + \nabla\cdot((\rho + p)\boldsymbol{\beta}) = 0. \tag{4.64}$$

Integrating this equation over a small three-dimensional volume surrounding the event and applying the three-dimensional Gauss's theorem, we see that the rate of change of the energy in the small volume equals the rate at which the velocity field is carrying energy into the volume plus the rate at which the pressure is doing work on the volume. Note again that this is the only flow of energy into the volume; the dynamics is describing a flow for which no heat is transmitted to a fluid element—an adiabatic process.

The space component of Eq. (4.63) can be written as

$$(\rho(x) + p(x))\mathbf{a} = -\nabla p(x), \qquad (4.65)$$

where $\mathbf{a}$ is the acceleration of the fluid element. Here $-\nabla p(x)$ is the force density exerted by the pressure. We see that the fluid in its rest frame has an "inertial mass density" that is equal to the $\rho(x) + p(x)$! Except for a very dense or very relativistic fluid, p is much less than ρ. For a nonrelativistic gas, $\rho \approx mn \equiv \rho_0$, where m is the mass of the particles and n is the number density. Thus, for a nonrelativistic gas, with p much less than ρ, $\rho_0(x)\mathbf{a} = -\nabla p(x)$.

If the fluid consists of pointlike, nonrelativistic, noninteracting particles, the energy of the nonrelativistic particles is $p^0 \approx m + m\beta^2/2$. (There is no interaction energy or energy associated with rotation or vibration of the particles.) For such a nonrelativistic ideal monatomic gas (see Exercise 9),

$$\rho \approx \rho_0 + \frac{3}{2}p. \qquad (4.66)$$

For an extremely relativistic gas of pointlike, noninteracting particles (see Exercise 9),

$$\rho \approx 3p. \qquad (4.67)$$

Such relationships, between the pressure and energy density, are referred to as *equations of state*.

The behavior of the fluid is then determined, in part, by the equation of state of the fluid and the conservation of energy-momentum equations. In addition, in some cases, the number of particles is conserved (i.e., there is no creation or destruction of particles.) For such cases, since a fluid element, defined as containing a fixed amount of particles, say N, undergoes only adiabatic changes, the relation between ρ, n and p is constrained, by the first law of thermodynamics, to satisfy

$$p\,d\left(\frac{N}{n}\right) + d\left(\frac{\rho N}{n}\right) = Td(sN) = 0.$$

Here T is the temperature and s is the entropy per particle. The adiabatic (constant entropy) constraint can then be written as

$$-p\frac{dn}{n} - \rho\frac{dn}{n} + d\rho = 0 \qquad (4.68)$$

or

$$\frac{\partial n}{\partial \rho}\bigg|_{s \, = \, constant} = \frac{n}{p + \rho}.$$ (4.69)

An important physical process that occurs in fluids, including relativistic fluids, is sound propagation. Sound propagation is governed by the dynamics resulting from the (local) conservation of the energy-momentum equation, with the velocity of sound in the rest frame of the fluid determined by the (adiabatic) equation of state. We can obtain the relation between the equation of state and the sound velocity by showing that energy density perturbations, in this frame, satisfy the wave equation. Consider, then, small perturbations $\delta p, \delta \rho$, and $\delta \boldsymbol{\beta}$ away from a uniform p_0 and ρ_0 and velocity zero, that is, in the rest frame of the fluid. To lowest order in $\delta p, \delta \rho$, and $\delta \boldsymbol{\beta}$, the time component of the conservation of the energy-momentum equation Eq. (4.64) becomes

$$\frac{\partial \delta \rho}{\partial x^0} + (\rho_0 + p_0) \nabla \cdot \delta \boldsymbol{\beta} = 0.$$ (4.70)

Similarly, for the space components, Eq. (4.65)

$$(\rho_0 + p_0) \mathbf{a} + \nabla \delta p (x) = 0.$$ (4.71)

With

$$\delta p (x) = \frac{\partial p}{\partial \rho}\bigg|_{\rho_0, \, s_0} \delta \rho,$$

Eq. (4.71) can be written

$$(\rho_0 + p_0) \mathbf{a} + (\frac{\partial p}{\partial \rho}\bigg|_{\rho_0, \, s_0}) \nabla \delta \rho (x) = 0.$$ (4.72)

Finally, by subtracting the time derivative of Eq. (4.70) from the divergence of Eq. (4.72), one obtains the wave equation satisfied by $\delta \rho$:

$$\frac{\partial^2 \delta \rho}{(\partial x^0)^2} - v_s^2 \nabla^2 \delta \rho = 0,$$ (4.73)

with the speed of sound, v_s, given by

$$v_s = (\frac{\partial p}{\partial \rho}\bigg|_{\rho_0, \, s_0})^{1/2}.$$ (4.74)

A clear physical constraint on the (adiabatic) equation of state is that the speed of sound is less than one. Note that, from Eq. (4.67), the speed of sound of relativistic gas is $(1/3)^{1/2}$. With the use of Eq. (4.69), for a nonrelativistic ideal gas, with an equation of state given by Eq. (4.66), one finds that $v_s \approx (5p/3mn)^{1/2}$, which is much smaller than the speed of light. (See Exercise 9.)

4.6.3   Energy-Momentum Tensor of the Electromagnetic Field

Elementary study of capacitors and inductors indicates that one can associate an energy density equal to $(\mathbf{E} \cdot \mathbf{E} + \mathbf{B} \cdot \mathbf{B})/8\pi$ for electric and magnetic fields. Similarly, a study of plane electromagnetic waves, shows that they carry a three-momentum density of $(\mathbf{E} \times \mathbf{B})/4\pi$. One might then expect that the energy-momentum tensor is quadratic in the fields. Thus, two contractions are needed to reduce the outer product of the two tensor fields to a field of rank 2. Two such terms are $n^{\alpha\beta} F_{\sigma\rho} F^{\sigma\rho}$ and $n_{\sigma\rho} F^{\alpha\sigma} F^{\rho\beta}$. There are other possible terms involving the dual tensor $\mathcal{F}^{\alpha\beta}$, Eq. (4.26). We do not use such terms but see if we can write a candidate energy-momentum tensor for the electromagnetic field as a linear combination of $n^{\alpha\beta} F_{\sigma\rho} F^{\sigma\rho}$ and $n_{\sigma\rho} F^{\alpha\sigma} F^{\rho\beta}$. Of course, in the absence of charges and currents the tensor must satisfy Eq. (4.52) and have an energy density, T^{00}, given by $(\mathbf{E} \cdot \mathbf{E} + \mathbf{B} \cdot \mathbf{B})/8\pi$. Such a tensor is

$$T_{EM}^{\alpha\beta} = \frac{1}{4\pi} \left(n_{\sigma\rho} F^{\alpha\sigma} F^{\rho\beta} + \frac{1}{4} n^{\alpha\beta} F_{\sigma\rho} F^{\sigma\rho} \right), \qquad (4.75)$$

which expressed in terms of the $\mathbf{E}$ and $\mathbf{B}$ fields becomes

$$T_{EM}^{\alpha\beta} = \frac{1}{4\pi} \begin{bmatrix} \frac{1}{2}(\mathbf{B} \cdot \mathbf{B} + \mathbf{E} \cdot \mathbf{E}) & B^3 E^2 - B^2 E^3 & -B^3 E^1 + B^1 E^3 & B^2 E^1 - B^1 E^2 \\ B^3 E^2 - B^2 E^3 & & & \\ -B^3 E^1 + B^1 E^3 & & 4\pi T^{ij} & \\ B^2 E^1 - B^1 E^2 & & & \end{bmatrix}$$

$$(4.76)$$

Here

$$T^{ij} = \frac{1}{4\pi} \left(-E^i E^j - B^i B^j + \frac{1}{2} \delta_{ij}(E^2 + B^2) \right).$$

We see that $T_{EM}^{00} = (\mathbf{B} \cdot \mathbf{B} + \mathbf{E} \cdot \mathbf{E})/8\pi$, the energy density, whereas $\mathbf{S} = (\mathbf{E} \times \mathbf{B})/4\pi$ is the momentum density as required.

We now show that, in the absence of charges and currents, the tensor $T_{EM}^{\alpha\beta}$ satisfies the local conservation law. Thus, from Eq. (4.75),

$$\partial_\alpha T^{\alpha\beta}_{EM} = \frac{1}{4\pi} \left[n_{\sigma\rho} \partial_\alpha F^{\alpha\sigma} F^{\rho\beta} + n_{\sigma\rho} F^{\alpha\sigma} \partial_\alpha F^{\rho\beta} + \frac{1}{2} n^{\alpha\beta} F_{\sigma\rho} \partial_\alpha F^{\sigma\rho} \right]. \quad (4.77)$$

The last two terms in the bracket can be written as

$$F_{\sigma\rho} \left[n^{\alpha\sigma} \partial_\alpha F^{\rho\beta} + \frac{1}{2} n^{\alpha\beta} \partial_\alpha F^{\sigma\rho} \right] = \frac{1}{2} F_{\sigma\rho} \left[\partial^\rho F^{\beta\sigma} + \partial^\sigma F^{\rho\beta} + \partial^\beta F^{\sigma\rho} \right]. \quad (4.78)$$

The factor in brackets on the right-hand side vanishes by Eq. (4.28). We then have

$$\partial_\alpha T^{\alpha\beta}_{EM} = \frac{1}{4\pi} n_{\sigma\rho} \partial_\alpha F^{\alpha\sigma} F^{\rho\beta} = J_\rho F^{\rho\beta} = -f^\beta. \quad (4.79)$$

Here f^β is the four-force density (see Eq. (4.33)). In a region where the four current vanishes, $T^{\alpha\beta}$ satisfies Eq. (4.52), as required.

4.6.4   Total Energy-Momentum Tensor of Charged Dust and Electromagnetic Field

Suppose the energy-momentum tensor of dust, Eq. (4.56), is not conserved; that is, it does not satisfy Eq. (4.52) because of interactions with other fields. It remains true that Eq. (4.56) still defines a tensor of rank 2 and thus $\partial_\beta T^{\alpha\beta}_D$ is a four-vector. In the rest frame of the dust at a particular event point, this vector is given by Eq. (4.58):

$$\partial_\beta T^{\alpha\beta}_D = \left(\frac{\partial \rho_0}{\partial x^0} + \rho_0(x) \nabla \cdot \boldsymbol{\beta} \right) \delta^{\alpha 0} + \rho_0(x) \frac{\partial \beta^j}{\partial x^0} \delta^{\alpha j}.$$

The time component still must vanish—the conservation of mass or number of particles. Even if some other field is interacting with the particles, there cannot be a change in the energy density due to this interaction since, in this frame, $\boldsymbol{\beta} = 0$. The space components are the rate of change of the three-momentum density, which must be equal to the three-force density. If the dust particles are charged, each with a charge q, the dust interacts with the electromagnetic field and the force density in this frame is ρE, where the charge density can be written $\rho = qn = (q/m)\rho_0 = (q/m)\rho_0 U^0$. In an arbitrary frame, the force density is given by the space part of

$$f^\alpha(x) = F^{\alpha\beta}(x) J_\beta = F^{\alpha\beta} \left(\frac{q}{m} \rho_0 U_\beta(x) \right).$$

Thus, for charged dust we have

$$\partial_\alpha T^{\alpha\beta}_D = f^\beta. \quad (4.80)$$

This, with Eq. (4.79), gives

$$\partial_\alpha (T_D^{\alpha\beta} + T_{EM}^{\alpha\beta}) = 0. \tag{4.81}$$

The sum of the energy-momentum of the charged dust and the electomagnetic field is locally conserved.

4.7   Exercises

1. Space inversion, $x'_i = -x_i$, satisfies Eq. (4.2), and the A that implements this transformation thus satisfies Eq. (4.3). (a) How does ϵ_{ijk} transform under space inversion? (b) What is the vector $\epsilon_{ijk} x_j p_k$ in Newtonian mechanics? How does it transform under space inversion?

2. (a) Space inversion, $x'^0 = x^0, x'^i = -x^i$, satisfies Eq. (4.10). How does $\epsilon^{\alpha\beta\rho\sigma}$ transform under space inversion? (b) Time inversion $x'^0 = -x^0, x'^i = x^i$ satisfies Eq. (4.10). How does $\epsilon^{\alpha\beta\rho\sigma}$ transform under time inversion?

3. Show that $\mathbf{E} \cdot \mathbf{B}$ and $\mathbf{E} \cdot \mathbf{E} - \mathbf{B} \cdot \mathbf{B}$ are scalar fields. (*Hint*: How can one form these scalars out of the electromagnetic field tensor?)

4. In a certain frame at a certain event point, $\mathbf{E} \cdot \mathbf{B} = 0$ and $\mathbf{E} \cdot \mathbf{E} < \mathbf{B} \cdot \mathbf{B}$. Prove there exists a frame in which $\mathbf{E} = 0$. Show that such a frame is not unique. (*Hint*: Choose a special "axis" and use a canonical Lorentz transformation.)

5. In a certain frame at a certain event, $\mathbf{E} \cdot \mathbf{B} \neq 0$. Prove that there exists a frame in which $\mathbf{E}$ and $\mathbf{B}$ are parallel or antiparallel. Show that such a frame is not unique.

6. (a) Transform, by a canonical Lorentz transformation, a general electromagnetic field tensor to the primed frame. (You are urged to use *Maple* or *Mathematica* to perform matrix multiplication to effect the Lorentz transformation.) (b) Show that the result can be written as

$$\mathbf{E}' = \gamma \mathbf{E} + \frac{\mathbf{E} \cdot \boldsymbol{\beta}}{\beta^2} (1 - \gamma) \boldsymbol{\beta} + \gamma \boldsymbol{\beta} \times \mathbf{B}$$

$$\mathbf{B}' = \gamma \mathbf{B} + \frac{\mathbf{B} \cdot \boldsymbol{\beta}}{\beta^2} (1 - \gamma) \boldsymbol{\beta} - \gamma \boldsymbol{\beta} \times \mathbf{E}.$$

One can use these three-vector relationships to give the connection between the fields as seen in frames with arbitrary relative velocity.

7. By transforming the field-strength tensor due to a charge q at rest in the primed frame, calculate the field in the unprimed frame in which the charge is moving in the x^1 direction with constant velocity β. Show that at the instant the charge is passing the origin

$$\mathbf{B} = \beta \times \mathbf{E}, \quad \mathbf{E} = \frac{q\mathbf{r}}{\gamma^2 r^3 \left[1 - \beta^2 \sin^2 \theta \right]^{3/2}},$$

where θ is the angle between the $\mathbf{r}$ and the x^1- axis. Note that the electric field and the magnetic field exist only in the $x^2 - x^3$ plane in the limit $\beta \to 1$. You might find the following identity useful: $\gamma^2 x^2 + y^2 + z^2 = \gamma^2 r^2 - (\gamma^2 - 1)(y^2 + z^2)$.

8. Consider a charged spinning particle, with a gyromagnetic ratio of two, that is accelerated by a magnetic field only. Show that

$$\frac{dS^0}{d\tau} = 0$$

$$\frac{dS \cdot \beta}{d\tau} = 0.$$

The latter result states that the component of the four-vector spin in the direction of β does not change. In turn, this implies in the rest frame of the particle that the component of spin in the direction tangent to the path doesn't change.

9. Let $n\left(\|\beta\|\right)$ be the number of molecules per unit volume with velocity components between β^i and $\beta^i + d\beta^i$. Thus, $\int n\left(\|\beta\|\right) d^3\beta = n$, where n is the number per unit volume. Recall that for a Newtonian ideal gas the pressure p is related to the average value of the kinetic energy of a molecule by

$$p = \frac{2}{3} n < \frac{m\beta^2}{2} >.$$

(a) By the usual argument, show more generally for a relativistic ideal gas that

$$p = \frac{1}{3} n < \mathbf{p} \cdot \beta >.$$

Here $<\mathbf{p} \cdot \beta> = (1/n) \int \mathbf{p} \cdot \beta n\left(\|\beta\|\right) d^3\beta$ is the average value of $\mathbf{p} \cdot \beta$.

(b) Derive the "equation of state" Eq. (4.66) for a nonrelativistic gas and (c) Eq. (4.67) for an extreme relativistic gas. (d) For such a nonrelativistic gas show that the velocity of sound is given by

$$v_s \approx (\frac{5p}{3mn})^{1/2}. \tag{4.82}$$

10. In a frame F, a ring, formed by a large number of particles of mass m rotates in the $x - y$ plane about the origin with an angular velocity ω. The ring has a radius r and a, small circular cross section of area δa. The particles, of number N, are uniformly distributed in the ring. (a) What is the number density in F? (b) What is the number density in the rest frame of the particles? (c) What is the energy-momentum tensor of this rotating "dust"? (d) What is the momentum flux at a given point in the torus? (e) How much momentum per unit time is passing a given cross section? (f) Using (e), how many particles per unit time are passing a given cross section? (g) Using (f), how many particles pass a given cross section in the time $2\pi r/r\omega$?

11. Use the energy momentum tensor $T^{\alpha\beta}$ of Exercise 10 to compute $\partial_\alpha T^{\alpha 0}$ and $\partial_\alpha T^{\alpha i}$. Are the results what you would expect for this tensor? Why? (Note that $\beta_x = -\omega y$ and $\beta_y = \omega x$.)

Chapter 5

Gravitation and Space-Time

5.1 Introduction

When Einstein published his special theory of relativity in 1905, there were two well-established forces of nature for which there was a theoretical description: the gravitational and the electromagnetic forces. Of course, as we have seen, the special theory of relativity fits well into the theoretical framework of electromagnetic theory—recall that Einstein's 1905 paper was titled "The Electrodynamics of Moving Bodies." Such is not the case with Newton's gravitational theory. Einstein's consideration of the gravitational force led to extensions of his ideas on the structure of space-time and culminated in that most beautiful of theories, his general theory of relativity. In this chapter we study ideas Einstein introduced in a paper published before his famous general theory of relativity paper. This study serves as a good transition for us—as it was perhaps for Einstein—between the special theory of relativity and the full-blown general theory.

5.2 Gravitation and Light

In 1911 Einstein published a paper titled "On the Influence of Gravitation on the Propagation of Light"[1] in which he attempted to calculate the effect of gravity on light propagation by applying an "equivalence principle," a principle by which the equality of gravitational mass and inertial mass is made manifest. Before we consider his

---

[1] *Annalen der Physik* 35 (1911), reprinted (in English translation) in Einstein et al. (1923).

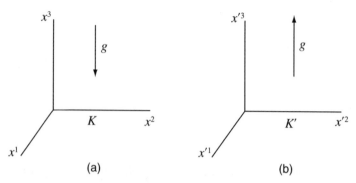

Figure 5.1. Equivalent frames.

arguments in this paper, let us recall some salient features of Newton's gravitational theory. Newton's law of universal gravitation states that the magnitude of the force of gravity that exists between two bodies of gravitational masses m and M separated by a distance r is given by

$$F = \frac{GmM}{r^2}. \tag{5.1}$$

Here G is Newton's universal gravitational constant. It was (and is) generally believed that the gravitational mass and the inertial mass of a body are the same. This equality implies that the acceleration g of a body of mass m due to the gravitational attraction of a body of mass M is independent of m and is given by

$$g = \frac{GM}{r^2}.$$

If time is measured in distance units, this becomes

$$g = \frac{GM}{c^2 r^2} = \frac{\tilde{G}M}{r^2}.$$

Here $\tilde{G} \equiv G/c^2$. One might look at this equality of gravitational and inertial mass as an accident—Einstein did not. He replaced this equality with a "hypothesis as to the nature of the gravitational field"[2] that we will now discuss following Einstein's original arguments.

Consider a stationary system of coordinates, K (Fig. 5.1a) in a region of space where there exists a homogeneous gravitational field with a gravitational acceleration g taken to be in the negative x^3 direc-

[2] Ibid.

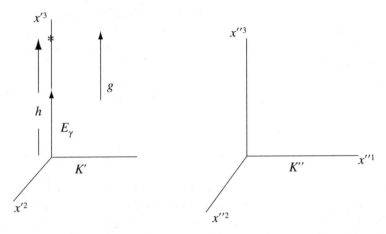

Figure 5.2. "Accelerated" photon.

tion. Consider also, in a region of space free of gravitational fields, a second system of coordinates K' (Fig. 5.1b) accelerating with a constant acceleration g in the positive x'^3 direction. If all relative velocities are small, so that the Galilean transformations of velocities apply, the motion of a particle free of forces other than gravity are the same in K and K'. Thus, as far as such motion is concerned, the two coordinate systems are equivalent. One cannot distinguish between the presence of a gravitational acceleration and an acceleration of the coordinate system. Einstein hypothesized a more general equivalence, namely, that the two coordinates are equivalent with respect to all physical processes and thus can be used to deduce the effects of gravity on light propagation.

We can infer what happens to a photon "rising" in coordinate system K by considering the photon in K'. In turn, we can conclude what happens in K' by considering the photon in a coordinate system K'' that is in the same region of space as K' but is not accelerating. (See Fig. 5.2.)

Let a photon of energy E_γ be emitted at the origin along the x'^3 axis at time $x'^0 = x''^0 = 0$ and received at a distance h from the origin. Since the K'' coordinate system is an inertial frame the photon proceeds up the x''^3 axis without changing energy. Assume K'' is at rest with respect to K' when the photon is emitted. For small relative velocity, the receiver's position in K'' as a function of x''^0 is

$$x''^3 \approx h + \frac{1}{2} g (x''^0)^2.$$

The photon is received in a time $x''^0 \approx x'^0 \approx h$. At that time K' is moving with velocity $\beta_r \approx gh \ll 1$ with respect to K''. When received, the photon has energy $E_{\gamma'}$, in an inertial frame comoving with the K' frame, given by

$$E'_\gamma \approx \gamma(E_\gamma - ghE_\gamma) \approx E_\gamma(1 - gh) \tag{5.2}$$

or

$$\frac{E'_\gamma - E_\gamma}{E_\gamma} = \frac{\delta E_\gamma}{E_\gamma} = -gh = -\delta\Phi. \tag{5.3}$$

$\delta\Phi$ is the change in the potential energy per unit mass in Newton's gravitational theory. From Einstein's hypothesis, this is also the fractional change of the photon's energy in a frame at rest in the presence of a gravitational acceleration g. If one uses the quantum condition $E_\gamma = h\nu$, Eq. (5.3) can be written

$$\frac{\delta\nu}{\nu} = -\delta\Phi. \tag{5.4}$$

This frequency shift can be argued from Einstein's equivalence hypothesis without invoking the quantum condition. Imagine that a source at the origin of the K' inertial coordinate system emits a light wave of fixed period T'_s in the x'^3 direction. In the K'' inertial frame the source moves up the x''^3 axis with a velocity $\beta_s \approx gx''^0$. The period of the wave emitted by the source, as measured in the K'' frame, is

$$T''_s = (1 - \beta_s^2)^{1/2} T'_s \approx (1 + \beta_s) T'_s \approx (1 + gx''^0) T'_s.$$

In the inertial frame K'' the period changes. We assume that g and T'_s are small enough so that there is very little change in β_s over many periods. We want to know what period the detector at $x'^3 = h$ measures. Since K'' is an inertial frame, the wave proceeds up the x''^3 axis with the velocity of light and with the (local) period unchanged. Thus, the light period, as observed in the K'' frame, at the detector at $x'^3 = h$ is the period of the light that was emitted at an earlier time $\Delta x''^0 \approx h$. With a knowledge of the period of the emitted light we can determine the period as observed in the K' frame using the Doppler shift to an inertial frame co-moving with the K' frame. We will use the inertial frame K'' that is moving with K' when the wave that is received at the detector is emitted. Thus, $T''_s = T'_s$. When this is received at $x'^3 = h \approx x''^3$, K' is moving with velocity $\beta_r \approx gh$. Thus, the period measured by the detector fixed in the K' frame, determined by the Doppler shift relation Eq. (2.9), is

$$T'_d = \frac{(1 + \beta_r)^{1/2}}{(1 - \beta_r)^{1/2}} T'_s \approx \frac{(1 + gh)^{1/2}}{(1 - gh)^{1/2}} T'_s \approx (1 + gh) T'_s.$$

With $\nu' = 1/T'_d$ and $\nu = T'_s$, this yields Eq. (5.4).

In passing, we note a hypothesis that one can use in place of that which Einstein assumed in his 1911 paper that is perhaps closer in spirit to the principle of equivalence of his later work and also results in Eq. (5.2). *The result of an observation made by a freely falling observer is the same as that of an observation made in an inertial frame in the absence of gravity.* The equivalence of this hypothesis to that of the 1911 paper is made manifest by viewing Figure 5.2 as depicting the freely falling frame as K'' and K' as the frame 'fixed' in the gravitational field, which of course is accelerating with respect to K''.

If one assumes that the expression for the change in frequency is valid even when the gravitational field is not homogeneous, but for which $\delta\Phi$ is still small, one could calculate the frequency shift for a photon emitted from the surface of a star of mass M and radius R and received far away ($r \approx \infty$). Then

$$\delta\Phi = \frac{\tilde{G}M}{R}.$$

For the sun $\delta\Phi \approx 2.12 \times 10^{-6}$ so that $\delta\nu/\nu \approx -2 \times 10^{-6}$. This change in frequency of one part in a million is difficult to detect. A Doppler shift caused by velocity of 0.6 km/sec would be of the same magnitude. Convective currents of hot gases on the surface of the sun easily exceed this velocity, thus effectively masking the *gravitational redshift* in frequency. It is referred to as a "redshift" because the frequency is shifted toward lower frequencies when the photon proceeds away from a massive body.

The gravitational shift was first observed on earth by groups at Harvard University and Harwell, England. In the Harvard experiment the photon "drops" (thus a blue shift) a distance of about 20 meters, resulting in a fractional frequency shift of about 10^{-14}, an incredibly small shift."[3]

5.3   Geometry Change in the Presence of Gravity

We can regard the frequency change that occurs when a photon moves into a region of different gravitational potential as an effect of

---

[3] See Pound and Rebka (1960).

Figure 5.3. Local wavelengths and local light cones.

a change of space-time geometry. These frequencies are those measured by clocks (i.e., "atomic clocks") at rest, located at the different positions. We refer to these clocks as *standard clocks*. The process we are considering is essentially a static process—the gravitational potential is not changing. Thus, if we consider the light as wave propagating between two points S_1 and S_2, there is a constant number of wavelengths between S_1 and S_2. Similarly, during the time, measured by either standard clock, a wave at S_1 goes through one cycle, and the wave at S_2 goes through one cycle. Otherwise there will be a loss or gain of the number of wavelengths between S_1 and S_2. Since the frequency at S_2 is different than at S_1, Eq. (5.4), as measured by standard clocks, the wavelength λ_2 at S_2 is different from the wavelength λ_1 at S_1. The relation between λ_1 and λ_2 is easily deduced by equating the velocity of light at S_2 (measured using a standard clock at S_2) with the velocity of light at S_1 (measured using a standard clock at S_1). Hence, we have

$$\nu_2 \lambda_2 = \nu_1 \lambda_1,$$

and, by using Eq. (5.4) (with $\nu = \nu_1, \delta \nu = \nu_2 - \nu_1$, and $\delta \Phi = \Phi_2 - \Phi_1$),

$$\nu_2 = \nu_1 [1 - (\Phi_2 - \Phi_1)]$$

we obtain

$$\lambda_2 = \frac{\lambda_1}{1 - (\Phi_2 - \Phi_1)}. \tag{5.5}$$

The wavelength increases as the wave goes "up."

In Figure 5.3 we have drawn this increase in wavelength for the case of $\Phi = gz$ and $gh \approx 1/2$. Of course $1/2$ is not "small" and, thus, our arguments for the shift are not valid, but we can at least depict the amount of the shift. We have drawn an x^0 axis, that is to be the time measured by a standard clock at S_1. This implies the light cone, at S_1, is a $45°$ cone (i.e., $c_1 = 1$).

However, since the wavelength at S_2 is twice that at S_1, the wave moves twice the distance in one period, that is, with twice the velocity (using the standard clock of S_1 for measuring time). Thus, the light cone at S_2 has a larger apex angle. In the absence of gravity the local light cones would be identical; the local light cone at any point could be obtained by merely translating the light cones from any other point. This is not true when gravitational fields are present.

Finally, we note that the velocity of light $c(\mathbf{x})$ at a point $\mathbf{x}$, as measured using a standard clock at the point $\mathbf{x}_0$, is given by

$$c(\mathbf{x}) = \lambda(\mathbf{x})\nu_0 = \frac{\lambda(\mathbf{x}_0)\nu_0}{1 - (\Phi(\mathbf{x}) - \Phi(\mathbf{x}_0))} \approx 1 + (\Phi(\mathbf{x}) - \Phi(\mathbf{x}_0)). \quad (5.6)$$

Here we have assumed $\Phi(\mathbf{x}) - \Phi(\mathbf{x}_0) \ll 1$. The geometry of light cones as depicted in Figure 5.3, with the velocity of light given by Eq. (5.6), is characterized by an invariant interval,

$$d\tau^2 = (1 + (\Phi(x) - \Phi(x_0)))^2 (dx^0)^2 - (dx^1)^2 - (dx^2)^2 - (dx^3)^2. \quad (5.7)$$

Here x^0 is the time as measured by a standard clock at point $\mathbf{x}_0$. It is clear the invariant interval represents a space-time geometry that differs from that of the special theory of relativity.

5.4 Deflection of Light in a Gravitational Field

Having argued that the velocity of light (measured by a fixed standard clock at one point in space) depends upon the position in the gravitational field, we can now calculate how much light is deflected when the rays pass through such a field by applying Huygen's principle. Recall that Huygen's principle states that to determine the position of wave front at some time $x^0 + \Delta x^0$ knowing what the wave front is at time x^0, one merely has to consider each point of the wave front at x^0 to be a source of spherical waves. That is, let these waves propagate for a time Δx^0, and the envelope of the spherical wave fronts form the wave front at $x^0 + \Delta x^0$.

Consider, then, a plane wave front of a light wave at time x^0. Let P_1 and P_2 be two points on the wave front separated by a distance δl where the velocities are c_1 and c_2 respectively. We see from Figure 5.4 that $\delta\phi$, the angle at which the wave is refracted in a time δx^0, is given by

$$\delta\phi \approx \frac{(c_1 - c_2)\delta x^0}{\delta l}.$$

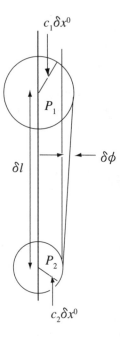

Figure 5.4. Bending of wave fronts.

In a gravitational field $c_1 - c_2 = \delta\Phi$, where $\delta\Phi$ is the change in the gravitational potential from point 1 to point 2. (See Eq. (5.6).) Thus, we have

$$\delta\phi = \frac{\delta\Phi}{\delta l}\,\delta x^0,$$

where $\delta\Phi/\delta l$ is the change per unit length of the gravitational potential along the wave front. It is also the component of the gravitational acceleration **g** along the wave front. Now consider a plane wave coming from far off and passing by a body of mass M, as depicted in Figure 5.5. The total deflection $\Delta\phi$ is given by

$$\Delta\phi = \int d\phi = \int \frac{d\Phi}{dl}\,dx^0.$$

Here the integral is taken over the full path, which is parameterized by x^0, the standard clock time at the position at which the gravitational potential has zero value. For light passing near the surface of the sun, the case that Einstein treated (and, of course, the case of most interest), the deflection calculated is extremely small. Thus, it is

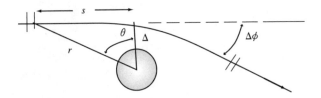

Figure 5.5. Gravitational deflection of a light ray.

a good approximation to merely sum up the infinitesimal deflections that would occur if the ray passed along the undeflected path. Similarly, the velocity of light along the path is very nearly constant so that $ds = dx^0$, where s measures the distance along the path. (See Fig 5.5.) Then we see that, when at a point characterized by (r, θ), in an elapsed time dx^0 (distance $ds = dx^0$), ϕ changes by

$$d\phi = \frac{d\Phi}{dl}\,ds = \frac{\tilde{G}M}{r^2}\cos(\theta)\,ds = \frac{\tilde{G}M\Delta ds}{(s^2 + \Delta^2)^{3/2}}.$$

Here Δ is the "impact parameter," which is also, approximately, the distance of closest approach to the mass M. The sum of $d\phi$ over the path, as the light comes in from far off to the left and proceeds out far off to the right, is the total expected deflection $\Delta\phi$:

$$\Delta\phi = \int d\phi = \int_{-\infty}^{\infty} \frac{\tilde{G}M\Delta ds}{(s^2 + \Delta^2)^{3/2}} = \frac{2\tilde{G}M}{\Delta}.$$

Using the mass of the sun for M and the radius of the sun for Δ, the point of closest approach, one obtains

$$\Delta\phi \approx 0.875 \text{ arcsec.}$$

This is the deflection that Einstein predicted in his 1911 paper for starlight that passes near the sun. Of course, usually one cannot observe starlight that passes near the sun—one would be observing in daylight. However, as Einstein noted in his paper, "As the fixed stars in the parts of the sky near the sun are visible during a total eclipse of the sun, this consequence of the theory may be compared to experiment."

We will see that this deflection is exactly half that predicted by Einstein's general relativity theory.

Chapter 6

General Relativity

6.1 Introduction

In his 1911 paper, Einstein deduced the effect of gravity on light by applying the principle of equivalence. However, the arguments were restricted to nonrelativistic relative velocities of inertial coordinates. He used Newton's theory of gravity in which effects are propagated instantly—the gravitational field due to a massive body depends only upon where it is at a given instant. This is incompatible with finite propagation of effects as implied by Einsteinian relativity. In addition, the principle of equivalence was applied over space-time regions for which the gravitational effects were small.

But the principle of equivalence is a very attractive idea and Einstein built a theory of gravity with this principle as a cornerstone.

In the previous chapter we argued that some of the results of the 1911 paper could be obtained by assuming a particular form of the space-time metric $g_{\alpha\beta}$ that is used to form the invariant interval

$$d\tau^2 = g_{\alpha\beta}\, dx^\alpha\, dx^\beta,$$

and that a world line of a photon is partially characterized by $d\tau^2 = 0$. Though there exist locally inertial coordinate systems, that is, coordinate systems corresponding to freely falling frames, for which

$$d\tau^2 = (dx^0)^2 - (dx^1)^2 - (dx^2)^2 - (dx^3)^2 = \eta_{\alpha\beta}\, dx^\alpha\, dx^\beta \qquad (6.1)$$

in some small space-time region around an event. However, a "global inertial coordinate" system did not exist, meaning there do not exist coordinates x^μ for which Eq. (6.1) is valid for all regions. With this observation in mind we can formulate the *principle of equivalence* as follows:

At every space-time point in a gravitational field it is possible to choose a "locally inertial coordinate system" such that within a sufficiently small region of the point the laws of physics take the same form as in an unaccelerated coordinate system in the absence of gravitation.

This principle implies that, given any point P in the four-dimensional space of events, one can find a set of four coordinates x^ν, whose origin is at P, in which the metric becomes locally Lorentzian; that is,

$$g_{\mu\nu}(x^\alpha) = \eta_{\mu\nu} + O((x^\alpha)^2).$$

Thus, in these coordinate systems, called *local Lorentz frames* or *local inertial frames*,

$$g_{\mu\nu}(P) = n_{\mu\nu} \tag{6.2}$$

$$\frac{\partial g_{\mu\nu}}{\partial x^\alpha}(P) = 0, \tag{6.3}$$

but generally

$$\frac{\partial^2 g_{\mu\nu}}{\partial x^\alpha \partial x^\beta}(P) \neq 0.$$

How to formally implement the principle of equivalence was the problem facing Einstein. A theory of gravity becomes a theory as to how the "local coordinate systems" are put together to form the geometry of space-time in the large: What is the space-time metric $g_{\alpha\beta}(x)$?

Einstein's theory of how local inertial coordinate systems are patched together was described in his paper titled "The Foundation of the General Theory of Relativity," published in 1916 (Einstein 1916). In this paper he developed equations for the space-time metric whose source was the energy-momentum distribution.

Generally, there does not exist a global coordinate system for which the space-time metric is independent of the event point. Thus, it is reasonable that one must consider general coordinate systems and the transformation of physical entities under general coordinate transformations. To assure that the *principle of equivalence* is satisfied, one can impose the *principle of general covariance*, which we take to consist of the following conditions (See Weinberg 1972, p. 72).

1. All physical equations obey the laws of special relativity in the absence of gravity.
2. The physical equations are *generally covariant*; that is, under a general change of coordinates the equations preserve their form.
3. The space-time metric is such that at any event point there exists a "locally inertial coordinate system."

The principle of general covariance implies the principle of equivalence. Requirement (2) implies that the physical equations are valid in locally inertial frames, and (1) implies that the equations in these frames obey the laws of special relativity in the absence of gravity. Requirement (2) gives the resulting theory its name, general relativity (GR). By itself, requirement (2) is somewhat empty. Many differential equations, which might be candidates for physical laws, can be written in generally covariant form by the inclusion of terms whose expression depends on the particular coordinate system used, terms that may in fact vanish in some particular coordinate system. It is the three conditions taken together that give content to the principle. In the following, I present, a somewhat heuristic derivation of the Einstein equations, based on the principle of general covariance.

In the study of special relativity we were led to the study of tensors, entities that transform in particular ways under Lorentz transformations. Special relativity invariance of equations was then assured if physical properties were expressed as tensors and physical equations were equations between like tensors. To study covariance under general coordinate transformations, one should understand tensors, entities that transform in particular ways under such general transformations.

In the following sections on tensors, geodesics, curvature, etc., the space emphasized is the four-dimensional space of events. The discussion can be generalized, both to spaces of arbitrary dimension N and to a slightly more general metric. When a "squared distance" on a space is defined by a quadratic differential form,

$$dl^2 = g_{ij}(x)\, dx^i dx^j, \tag{6.4}$$

such that $dl^2 \geq 0$ and such that $dl^2 = 0$ implies that $dx^i = 0$, then the space is referred to as a *Riemannian space*. The metric is said to be positive definite. A *local frame theorem* states that at any point P in the space one can choose coordinates such that

$$g_{ij}(P) = \delta_{ij} \tag{6.5}$$

$$\frac{\partial g_{ij}}{\partial x^i}(P) = 0, \tag{6.6}$$

that is, there exist coordinates such that the metric is locally Euclidean. The metric induced on a two-sphere by the imbedding in the three-dimensional flat Euclidean space is an example of such a Riemannian metric:

$$dl^2 = +r^2\sin^2(\theta)\,d\phi^2 + r^2\,d\theta^2. \tag{6.7}$$

If the condition that the metric be positive definite is relaxed to the requirement that it be nondegenerate, the space (or metric) is said to be *pseudo-Riemannian*. "Non-degenerate" simply means that $g_{\alpha\beta}V^\beta = 0$ implies $V^\beta = 0$. That is, treated as a matrix, $\mathbf{g}$ has no zero eigenvalue. For such a case, the local frame theorem states that at any point in the space one can choose coordinates such that

$$g_{\mu\nu}(P) = n_{\mu\nu} \tag{6.8}$$

$$\frac{\partial g_{\mu\nu}}{\partial x^\alpha}(P) = 0, \tag{6.9}$$

where $n_{\nu\mu}$ is a diagonal matrix with $+1$'s and -1's on the diagonal. The number of $+1$'s and -1's on this diagonal matrix is characteristic of the metric and is referred to as the *signature* of the metric. The space-time metric is an example of a pseudo-Riemannian metric with but one $+1$ and is referred to as a Lorentzian metric.

6.2   Tensors of General Coordinate Transformations

We need to consider tensors whose components can be defined with respect to an arbitrary system of coordinates that label a particular event (point) in our space. As in the case of tensors of special relativity, the archetype of a *contravariant* rank 1 tensor is a differential of event space dx^μ. The (four) coordinates x^μ are not necessarily the space-time of a standard clock and an orthogonal lattice. In fact, such global coordinates in general do not exist. The transformation of these differentials from x^μ to coordinates $x'^\nu(x)$ is given by the chain rule,

$$dx'^\nu = \frac{\partial x'^\nu}{\partial x^\mu}\,dx^\mu. \tag{6.10}$$

(The transformations of the differential of space-time coordinates under Lorentz transformations are a particular subset of such transformations.) The transformation of any contravariant rank 1 tensor $A^\mu(x)$ is given by

$$A'^\nu(x') = \frac{\partial x'^\nu}{\partial x^\mu} A^\mu(x). \tag{6.11}$$

As under Lorentz transformations, the archetype of a *covariant* vector of rank 1 is the gradient of a scalar field $\partial_\mu \theta(x)$. Recall that a scalar field satisfies $\theta'(x') = \theta(x)$. The chain rule in the form

$$\partial_{\nu'} = \frac{\partial x^\mu}{\partial x'^\nu} \partial_\mu$$

implies

$$\partial_{\nu'}\theta'(x') = \frac{\partial x^\mu}{\partial x'^\nu} \partial_\mu \theta(x). \tag{6.12}$$

The primed index on the symbol $\partial_{\nu'}$ indicates a derivative with respect to a primed coordinate.

The meaning of mixed tensor of contravariant rank n and covariant rank m, defined by its transformation, is clear:

$$A'^\nu_\alpha(x') = \frac{\partial x^\beta}{\partial x'^\alpha} \frac{\partial x'^\nu}{\partial x^\mu} A^\mu_\beta(x). \tag{6.13}$$

As for Lorentz tensors, the *outer product* of two tensors, for example $A_{\rho\beta} B^\sigma$, forms a tensor of ranks that are the sums of the ranks of the two tensors. Similarly, it is easily shown that the operation of *contraction* of a covariant index with a contravariant index produces a tensor of one lower contravariant rank and one lower covariant rank.

Clearly, it is important to know how the metric $g_{\alpha\beta}(x)$ transforms. Its transformation is implied by the scalar character of the (local) invariant interval. That is, $d\tau^2 = g_{\alpha\beta}(x)\,dx^\alpha\,dx^\beta = g'_{\nu\mu}(x')\,dx'^\nu\,dx'^\mu$, or

$$d\tau^2 = g_{\alpha\beta}(x)\,dx^\alpha\,dx^\beta = g_{\alpha\beta}(x)\left(\frac{\partial x^\alpha}{\partial x'^\nu}dx'^\nu\right)\left(\frac{\partial x^\beta}{\partial x'^\mu}dx'^\mu\right) = g'_{\nu\mu}(x')\,dx'^\nu\,dx'^\mu.$$

We see that

$$g'_{\nu\mu}(x') = g_{\alpha\beta}(x)\frac{\partial x^\alpha}{\partial x'^\nu}\frac{\partial x^\beta}{\partial x'^\mu}. \tag{6.14}$$

The metric transforms like a tensor of covariant rank 2. This might have been expected since $dx^\alpha\,dx^\beta$ transforms as a tensor of contravariant rank 2 and to form a scalar, $d\tau^2$, we would expect these contravariant indices to be contracted with two covariant indices.

As for Lorentz tensors, the "inverse" metric $g^{\alpha\beta}$, defined by $g^{\alpha\beta} g_{\beta\rho} = \delta^\alpha_\rho$, is a contravariant rank 2 tensor, for we have

$$g^{\alpha\beta}(x) \frac{\partial x'^{\mu}}{\partial x^{\beta}} \frac{\partial x'^{\sigma}}{\partial x^{\alpha}} g'_{\mu\nu}(x') = g^{\alpha\beta}(x) \frac{\partial x'^{\mu}}{\partial x^{\beta}} \frac{\partial x'^{\sigma}}{\partial x^{\alpha}} \frac{\partial x^{\rho}}{\partial x'^{\mu}} \frac{\partial x^{\eta}}{\partial x'^{\nu}} g_{\rho\eta}$$

$$= g^{\alpha\beta}(x) \frac{\partial x'^{\sigma}}{\partial x^{\alpha}} \frac{\partial x^{\eta}}{\partial x'^{\nu}} \delta^{\rho}_{\beta} g_{\rho\eta}$$

$$= \delta^{\alpha}_{\eta} \frac{\partial x'^{\sigma}}{\partial x^{\alpha}} \frac{\partial x^{\eta}}{\partial x'^{\nu}} = \delta^{\sigma}_{\nu}.$$

6.3 Path of Freely Falling Particles: Timelike Geodesics

Differentiation of a Lorentz tensor yields a Lorentz tensor of one higher covariant rank. However, differentiation of these more general tensors does not result in a tensor because the coordinate transformations are not restricted to be linear like Lorentz transformations. Before investigating how one might define a "differentiation" operation on general tensors that will result in a tensor, we will consider a particular example that we might expect to result in a quantity that has tensorial transformation properties. Thus, we study the equation satisfied by the motion of a particle that is moving under the influence of gravity only—a freely falling particle. For any event on the world line of the particle there exists a locally inertial frame, with coordinates ζ^{μ}, for which

$$g_{\mu\nu}(\zeta) = n_{\mu\nu} + O(\zeta^2).$$

By the principle of general covariance, the equation satisfied by the particle at $\zeta = 0$ is

$$\frac{d^2 \zeta^{\mu}}{d\tau^2} = 0. \tag{6.15}$$

In local inertial coordinates the world line is a straight line if one restricts to a small enough region about $\zeta = 0$. Here $d\tau$ is the invariant interval along the world line of the particle. Note that there exists more than one inertial frame at a point, for if $g_{\mu\nu}(\zeta) = n_{\mu\nu} + O(\zeta^2)$, a change of coordinates $\zeta'^{\nu} = \Lambda^{\nu'}_{\mu} \zeta^{\mu} + O(\zeta^2)$ gives $g'_{\mu\nu} = n_{\mu\nu} + O(\zeta'^2)$, if $\Lambda^{\nu'}_{\mu}$ is a Lorentz transformation. We are parameterizing the world line with the proper time measured along the world line, a parameterization that must be modified when considering the motion of a particle moving with the velocity of light. Eq. (6.15) is locally that of a timelike geodesic, as we know that a free particle follows a path of longest proper time between two points on its worldline. (There are paths that in local coordinates are straight lines but which are spacelike, which we call *spacelike geodesics.*) Let us consider the form that Eq. (6.15) takes in an arbitrary coordinate system x to which the local coordinate system ζ is related by the transformation $\zeta^\mu(x^\alpha)$. Writing $\zeta^\alpha(x(\tau))$, we have

$$\frac{d\zeta^\alpha}{d\tau} = \frac{\partial \zeta^\alpha}{\partial x^\beta} \frac{dx^\beta}{d\tau}, \tag{6.16}$$

from which

$$\frac{d^2 \zeta^\mu}{d\tau^2} = \frac{d^2 x^\beta}{d\tau^2} \frac{\partial \zeta^\alpha}{\partial x^\beta} + \frac{\partial^2 \zeta^\alpha}{\partial x^\rho \partial x^\beta} \frac{dx^\beta}{d\tau} \frac{dx^\rho}{d\tau} = 0. \tag{6.17}$$

Multiplying Eq. (6.17) by $\partial x^\nu / \partial \zeta^\alpha$ and summing over the repeated index α, we obtain

$$\frac{d^2 x^\nu}{d\tau^2} + \frac{\partial^2 \zeta^\alpha}{\partial x^\rho \partial x^\beta} \frac{\partial x^\nu}{\partial \zeta^\alpha} \frac{dx^\beta}{d\tau} \frac{dx^\rho}{d\tau} = 0. \tag{6.18}$$

With a definition of *Christoffel symbols* $\Gamma^\nu_{\rho\beta}(x)$ as

$$\Gamma^\nu_{\rho\beta}(x) = \frac{\partial^2 \zeta^\alpha}{\partial x^\rho \partial x^\beta} \frac{\partial x^\nu}{\partial \zeta^\alpha}, \tag{6.19}$$

Eq. (6.18) can be written

$$\frac{d^2 x^\nu}{d\tau^2} + \Gamma^\nu_{\rho\beta}(x) \frac{dx^\beta}{d\tau} \frac{dx^\rho}{d\tau} = 0. \tag{6.20}$$

Thus, we have the equation for a freely falling particle in arbitrary coordinates but the equation involves the Christoffel symbols that are expressed in terms of the transformation equations between local inertial coordinates and the general coordinates. Eq. (6.20) is an equation for a geodesic, a path in space-time that has the longest proper time between any two points on the path, and thus one expects that the equation can be expressed in terms of the metric for arbitrary coordinates. (Note that under a change of parameterization of the path from τ to $\tilde{\tau} = a\tau$, with a a constant, the geodesic equation is unchanged. Such parameters are called *affine parameters.*) We now show that $\Gamma^\nu_{\rho\beta}(x)$ can be expressed in terms of the metric. First, note that the transformation property of the metric implies that, near an event point for which ζ^β are local inertial coordinates,

$$g_{\mu\nu}(x) = \frac{\partial \zeta^\alpha}{\partial x^\mu} \frac{\partial \zeta^\beta}{\partial x^\nu} g_{\alpha\beta}(\zeta) = \frac{\partial \zeta^\alpha}{\partial x^\mu} \frac{\partial \zeta^\beta}{\partial x^\nu} (\eta_{\alpha\beta} + O(\zeta^2)). \tag{6.21}$$

The derivative of Eq. (6.21) with respect to x^σ, evaluated at $\zeta = 0$, gives

$$\frac{\partial g_{\mu\nu}}{\partial x^\sigma}(x) = n_{\alpha\gamma}\left[\frac{\partial^2 \zeta^\alpha}{\partial x^\sigma \partial x^\mu}\frac{\partial \zeta^\gamma}{\partial x^\nu} + \frac{\partial^2 \zeta^\gamma}{\partial x^\sigma \partial x^\rho}\frac{\partial \zeta^\alpha}{\partial x^\mu}\right]. \qquad (6.22)$$

From Eq. (6.19) we have

$$\frac{\partial^2 \zeta^\alpha}{\partial x^\rho \partial x^\beta} = \Gamma^\nu_{\rho\beta}(x)\frac{\partial \zeta^\alpha}{\partial x^\nu}. \qquad (6.23)$$

Using this and Eq. (6.21) in Eq. (6.22), we find

$$\frac{\partial g_{\mu\nu}}{\partial x^\sigma} = \Gamma^\eta_{\sigma\mu}g_{\eta\nu} + \Gamma^\eta_{\sigma\nu}g_{\eta\mu}. \qquad (6.24)$$

Now add to Eq. (6.24) the same equation with μ interchanged with σ, and subtract the same equation with ν interchanged with σ to obtain

$$\frac{\partial g_{\mu\nu}}{\partial x^\sigma} + \frac{\partial g_{\sigma\nu}}{\partial x^\mu} - \frac{\partial g_{\mu\sigma}}{\partial x^\mu} = 2\Gamma^\eta_{\sigma\mu}g_{\eta\nu}. \qquad (6.25)$$

We have used the fact that $\Gamma^\nu_{\alpha\beta}$ is symmetric under interchange of α and β. By multiplying the above equation with $g^{\nu\rho}$, we have

$$\Gamma^\rho_{\sigma\mu} = \frac{1}{2}g^{\nu\rho}\left[\frac{\partial g_{\mu\nu}}{\partial x^\sigma} + \frac{\partial g_{\sigma\nu}}{\partial x^\mu} - \frac{\partial g_{\mu\sigma}}{\partial x^\nu}\right]. \qquad (6.26)$$

Having succeeded in expressing the Christoffel symbols in terms of the metric and its derivative evaluated in any coordinate system, we can write the equation for a freely falling particle, Eq. (6.20), in terms of the metric in any coordinate system. In a locally inertial coordinate system, $\Gamma^\rho_{\alpha\beta} = 0$ and Eq. (6.20) reduces to Eq. (6.15).

Finally, note that the geodesics of Riemannian spaces are paths of the shortest distance between two points, such as the great circles on a two-dimensional sphere, in contrast to timelike geodesics for Lorentzian metrics that have the longest proper time between events. There is a caveat that should be added to the last statement: timelike geodesics have a longer proper time than any "nearby" path between two events.

6.4   Covariant Differentiation

Since Eq. (6.20) is valid in any coordinate system, one might expect that, even though the two terms in the equation may not transform separately as a contravariant vector of rank 1, the sum may. We can use this observation to guess how to define a "differentiation" that might result in tensors. First note that $dx^\nu/d\tau$ is a vector since

$$\frac{dx'^{\mu}}{d\tau} = \frac{\partial x'^{\mu}}{\partial x^{\nu}} \frac{dx^{\nu}}{d\tau} .$$

With $A^{\nu} = dx^{\nu}/d\tau$, Eq. (6.20) can be written as

$$\frac{dA^{\nu}}{d\tau} + \Gamma^{\nu}_{\rho\beta}(x) A^{\beta} \frac{dx^{\rho}}{d\tau} = 0.$$

This suggests that for any contravariant vector $A^{\nu}(x)$, the object $DA^{\nu}/d\tau$ defined by

$$\frac{DA^{\nu}}{d\tau} = \frac{dA^{\nu}}{d\tau} + \Gamma^{\nu}_{\rho\beta}(x) A^{\beta} \frac{dx^{\rho}}{d\tau} \qquad (6.27)$$

is a contravariant vector. One can check that it is. It is called the *directional absolute derivative*, or the *directional covariant derivative*, in the direction $dx^{\rho}/d\tau$. For any parameterized curve $x^{\rho}(s)$,

$$\frac{DA^{\nu}}{ds} = \frac{dA^{\nu}}{ds} + \Gamma^{\nu}_{\rho\beta}(x) A^{\beta} \frac{dx^{\rho}}{ds} \qquad (6.28)$$

is a tensor, the absolute derivative of A^{ν} in the direction of dx^{ρ}/ds.
  Furthermore, since

$$\frac{dA^{\nu}}{ds} = \frac{\partial A^{\nu}}{\partial x^{\rho}} \frac{dx^{\rho}}{ds},$$

we can write

$$\frac{DA^{\nu}}{ds} = \left[\frac{\partial A^{\nu}}{\partial x^{\rho}} + \Gamma^{\nu}_{\rho\beta}(x) A^{\beta} \right] \frac{dx^{\rho}}{ds},$$

which suggests that

$$A^{\nu}_{;\rho} \equiv \frac{\partial A^{\nu}}{\partial x^{\rho}} + \Gamma^{\nu}_{\rho\beta}(x) A^{\beta} \qquad (6.29)$$

is a mixed tensor of covariant and contravariant rank 1. One can show that it is. It is called the *covariant derivative* of $A(x)$ (note the notation "; ρ" for the covariant derivative). The connection between the absolute derivative DA^{ν}/ds and the covariant derivative $A^{\nu}_{;\rho}$ is

$$\frac{DA^{\nu}}{ds} = A^{\nu}_{;\rho} \frac{dx^{\rho}}{ds} . \qquad (6.30)$$

  Of course, the equation for a geodesic that describes a freely falling particle, Eq. (6.20), can be written in terms of a covariant derivative. After all, we started our argument to suggest a definition of a covariant derivative with Eq. (6.20). First, note that the covariant definition of the four-velocity U^{μ} is, of course, $U^{\mu} = dx^{\mu}/d\tau$; this definition reduces to the special relativity definition in a locally inertial frame. By use of the four-velocity, Eq. (6.20) can be written as

$$\frac{DU^\nu}{d\tau} = U^\nu_{;\rho} U^\rho = 0. \tag{6.31}$$

Furthermore, by use of the covariant definition of the four-momentum, $p^\mu = mU^\mu$, Eq. (6.31) can be written as

$$\frac{Dp^\mu}{d\tau} = 0, \tag{6.32}$$

which in a locally inertial coordinate system reduces to $dp^\mu/d\tau = 0$.

The covariant derivative of an arbitrary tensor $A^{\beta\cdots}_{\alpha\cdots}$ is given by

$$A^{\beta\cdots}_{\alpha\cdots;\rho} = A^{\beta\cdots}_{\alpha\cdots,\rho} + \Gamma^\beta_{\sigma\rho} A^{\sigma\cdots}_{\alpha\cdots} + \ldots - \Gamma^\sigma_{\alpha\rho} A^{\beta\cdots}_{\sigma\cdots} - \ldots \tag{6.33}$$

Here we have introduced the notation ";ρ" for "$\partial/\partial x^\rho$."

It is important to note that $g^{\alpha\beta}_{;\rho} = 0$ and $g_{\alpha\beta;\rho} = 0$. These follow easily from the fact that tensor equations that are true in a local inertial coordinate system are satisfied in all coordinate systems. $g^{\alpha\beta}$ and $g_{\alpha\beta}$ can be taken in and out of covariant differentiation.

6.5   Parallel Transport: Curvature Tensor

Consider some path $x^\mu(s)$ in event space parameterized by s and a contravariant vector A^ν that satisfies

$$\frac{DA^\nu}{ds} = \frac{dA^\nu}{ds} + \Gamma^\nu_{\rho\beta}(x) A^\beta \frac{dx^\rho}{ds} = 0,$$

or

$$\frac{dA^\nu}{ds} = -\Gamma^\nu_{\rho\beta}(x) A^\beta \frac{dx^\rho}{ds}. \tag{6.34}$$

By considering this equation in a locally inertial frame, we can give it a geometric interpretation. In such a frame, since $\Gamma^\nu_{\rho\beta} = 0$, Eq. (6.34) becomes

$$\frac{dA^\nu}{ds} = 0.$$

That is, the components of the vector, in a locally inertial frame, do not change as the parameter changes—the vector is parallel-transported along the path. Note that if A^ν satisfies Eq. (6.34), then

$$\frac{dA^\nu A_\nu}{ds} = \frac{DA^\nu g_{\nu\mu} A^\mu}{ds} = 0. \tag{6.35}$$

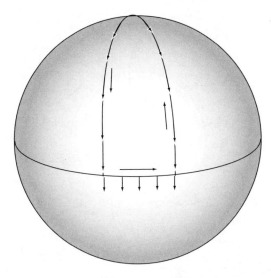

Figure 6.1. Parallel transport on a sphere.

This follows from $g_{\alpha\beta;\rho} = 0$. Similarly, if the path is a geodesic path, that is, it satisfies Eq. (6.31), then

$$\frac{dU^{\alpha} A_{\alpha}}{d\tau} = 0.$$

The scalar $U^{\alpha} A_{\alpha}$ is constant along the path. Loosely speaking, the "component" of A in the direction of the path doesn't change.

A natural question to ask is, "If a vector, starting at a given point, is parallel-transported around a closed path, thus returning to the starting point, does the original vector result?" Consider parallel-transporting a two-dimensional vector on the surface of the sphere around a close path starting at a pole, proceeding along a fixed longitude to the equator, following the equator to some point, and then move back to the pole along a fixed longitude, as depicted in Figure 6.1. The path chosen is a sequence of path sections each of which is a geodesic of the sphere. Thus, the angle the vector makes with each path section does not change as the vector is transported along the section . We see that the resulting vector is generally not the original vector, but has been rotated by an angle equal to the angle made by the intersection of the two longitudes. The direction of rotation (counterclockwise) is the same as the path direction. Of course, if the two-dimensional surface is a flat surface, parallel transport of a vector around a closed path does not change the vector. A change in the vector results if the surface is curved.

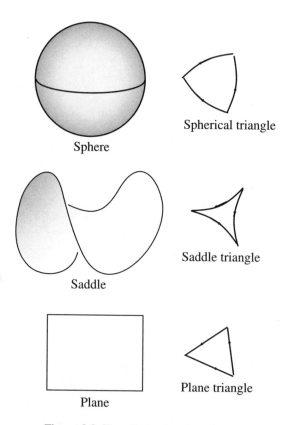

Figure 6.2. Two-dimensional surfaces.

But the characterizations of surfaces as being "flat" or "spherical" in a region is not exhaustive. Consider a two-dimensional surface that curves "up" in one direction and "down" in a perpendicular direction—a "saddle" surface. Such a surface is depicted with a spherical surface and a plane (flat) surface in Figure 6.2. Note that the intrinsic geometric properties of spherical and plane surfaces are everywhere the same. Not so for the saddle surface—the point at the "center" of the saddle is the most symmetric point. Also depicted in the figure are small "triangles" formed by geodesics for the three surfaces. The triangle for the saddle is centered at the center of the saddle. Of course, for the sphere and saddle triangle, the sides do not all lie in the same two-dimensional plane of the embedding three-space. Nevertheless, we see that the sum of the three angles made by the intersections of the geodesics is $>\pi$ for the sphere, $<\pi$ for the saddle, and $=\pi$ for the plane surface. One can imagine that parallel-

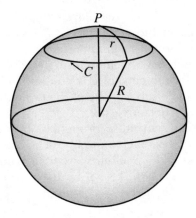

Figure 6.3. Curvature of sphere.

transporting a vector around the spherical triangle in a counter-clockwise direction as indicated in the figure results in a rotation of the vector in a counterclockwise direction, whereas for the saddle triangle the vector would be rotated in clockwise direction. Of course, for the plane triangle the vector is not rotated. The sphere is said to have positive "curvature," the saddle negative "curvature," and the plane zero "curvature."

To obtain a quantitative definition of curvature at a point P of a two-dimensional surface that is intrinsic to the surface, consider the distance C along a closed path that is the locus of all points that are at the same geodesic distance r from the point. For the case of the sphere (Fig. 6.3) for small r/R,,

$$C = 2\pi R \sin\left(\frac{r}{R}\right) \approx 2\pi \left(r - \frac{r^3}{6R^2}\right) = 2\pi \left(r - \frac{\kappa r^3}{6}\right).$$

Here $\kappa = 1/R^2$, a reasonable definition for the curvature of a sphere. From this we obtain the following formula for this curvature:

$$\kappa = \frac{3}{\pi} \lim_{r \to 0} \frac{2\pi r - C}{r^3}, \tag{6.36}$$

a result that depends only on the intrinsic properties of the metric of the two-dimensional surface with no reference to the embedding space. The sign of the curvature is positive if $C < 2\pi r$, negative if $C > 2\pi r$, and zero if $C = 2\pi r$. With Eq. (6.36) as the definition of the curvature at a point P of any two-dimensional surface, then

$$\kappa = \frac{1}{R_1 R_2} \tag{6.37}$$

obtains, where R_1 and R_2 are the extrema of the radii of sections formed by planes normal to the surface at P with *appropriate sign*. (A section is the curve formed by the intersection of the surface and a plane.) If the two extrema, which occur in perpendicular directions, are of the same sign, the curvature is positive, and if opposite it is negative. For the saddle surface, one of the two radii is negative and thus it has a negative curvature.

To characterize the curvature properties in a small space-time region, we study the parallel transport of a contravariant vector around a "small" closed loop in our space-time manifold. The closed path is given by $x^\mu(s)$, with $0 \le s \le 1, x^\mu(0) = x^\mu(1) = x_0^\mu$. From Eq. (6.34) the change ΔA^ν in a vector A^μ is given by

$$\Delta A^\nu = \int_0^1 \frac{dA^\nu}{ds} ds = - \int_0^1 \Gamma^\nu_{\rho\beta}(x) A^\beta(x) \frac{dx^\rho}{ds} ds. \qquad (6.38)$$

Expanding to lowest order around x_0, we have

$$\Gamma^\nu_{\rho\beta}(x) \approx \Gamma^\nu_{\rho\beta}(x_0) + \frac{\partial \Gamma^\nu_{\rho\beta}(x_0)}{\partial x^\alpha} (x^\alpha - x_0^\alpha), \qquad (6.39)$$

and by use of Eq. (6.34),

$$A^\beta(x) \approx A^\beta(x_0) - \Gamma^\beta_{\alpha\sigma}(x_0) A^\sigma(x_0) [x^\alpha - x_0^\alpha]. \qquad (6.40)$$

Using these last two approximations in Eq. (6.38) one obtains

$$\Delta A^\nu = - \int_0^1 \left[\Gamma^\nu_{\rho\beta}(x_0) + \frac{\partial \Gamma^\nu_{\rho\beta}(x_0)}{\partial x^\alpha} (x^\alpha - x_0^\alpha) \right]$$
$$\left[A^\beta(x_0) - \Gamma^\beta_{\alpha\sigma}(x_0) A^\sigma(x_0) (x^\alpha - x_0^\alpha) \right] \frac{dx^\rho}{ds} ds.$$

Note that dx^ρ/ds is "small" and thus there is no zero-order contribution to ΔA^ν, and since

$$\int_0^1 \frac{dx^\rho}{ds} ds = 0,$$

the first-order contribution vanishes. We see then, to second order, that

$$\Delta A^\nu = - \left[\frac{\partial \Gamma^\nu_{\rho\beta}(x_0)}{\partial x^\alpha} - \Gamma^\nu_{\rho\eta}(x_0) \Gamma^\eta_{\alpha\beta}(x_0) \right] A^\beta(x_0) \int_0^1 [x^\alpha - x_0^\alpha] \frac{dx^\rho}{ds} ds.$$

The integrals are antisymmetric in their indices. This can be seen by integrating by parts as follows:

$$\int_0^1 [x^\alpha - x_0^\alpha] \frac{dx^\rho}{ds} ds = [x^\alpha - x_0^\alpha] x^\rho |_{s=0}^{s=1} - \int_0^1 \frac{d[x^\alpha - x_0^\alpha]}{ds} x^\rho ds$$

$$= 0 - \int_0^1 \frac{dx^\alpha}{ds} [x^\rho - x_0^\rho] ds.$$

From this it follows that

$$\Delta A^\nu = \left[\frac{\partial \Gamma_{\beta\alpha}^\nu (x_0)}{\partial x^\rho} - \frac{\partial \Gamma_{\beta\rho}^\nu (x_0)}{\partial x^\alpha} + \Gamma_{\rho\eta}^\nu (x_0) \Gamma_{\beta\alpha}^\eta (x_0) - \Gamma_{\alpha\eta}^\nu (x_0) \Gamma_{\beta\rho}^\eta (x_0) \right]$$

$$A^\beta (x_0) \left[\frac{1}{2} \int_0^1 [x^\alpha - x_0^\alpha] \frac{dx^\rho}{ds} ds \right]. \tag{6.41}$$

Since ΔA^ν is the difference of two vectors *at the same event point*, it is a vector, as is A^β. Similarly the integral is a tensor of contravariant rank 2. It consists of a sum of the product of two "small" changes in event points. Thus, we might expect that $R_{\beta\alpha\rho}^\nu$, defined by

$$R_{\beta\alpha\rho}^\nu = \Gamma_{\beta\alpha, \rho}^\nu - \Gamma_{\beta\rho, \alpha}^\nu + \Gamma_{\rho\eta}^\nu \Gamma_{\beta\alpha}^\eta - \Gamma_{\alpha\eta}^\nu \Gamma_{\beta\rho}^\eta, \tag{6.42}$$

is a mixed tensor of contravariant rank 1 and covariant rank 3. One can show it is. It is called the *Riemann curvature tensor*. If at a given space-time point, there exists a coordinate system such that $g_{\alpha\beta}(\zeta) = n_{\alpha\beta} + O(\zeta^3)$, then $R_{\beta\alpha\rho}^\nu = 0$.

We were led to the Riemann curvature tensor by considering parallel-transporting vectors around infinitesimal closed paths. By considering a closed path defined by the sequence of four infinitesimals $da^\beta, db^\sigma, - da^\beta, - db^\sigma$, one might expect that the Riemann tensor would be involved if one considered the action on a vector field V^α of the "commutator" of the two covariant derivatives in the directions da^β and db^σ. "Commutator" means the difference between the actions of taking the derivatives first in one order and then reversing the order. That is,

$$V_{;\beta;\sigma}^\alpha - V_{;\sigma;\beta}^\alpha.$$

This commutator is a tensor field of contravariant rank 1 and covariant rank 2. In a locally inertial frame, for which $\Gamma_{\beta\rho}^\alpha = 0$ but for which $\Gamma_{\beta\rho, \sigma}^\alpha$ generally does not vanish, this commutator is easy to compute with the following result:

$$V_{;\beta;\sigma}^\alpha - V_{;\sigma;\beta}^\alpha = (\Gamma_{\rho\beta, \sigma}^\alpha - \Gamma_{\rho\sigma, \beta}^\alpha) V^\rho.$$

But, since in a local inertial frame $R_{\beta\rho\sigma}^\alpha = \Gamma_{\beta\rho, \sigma}^\alpha - \Gamma_{\beta\sigma, \rho}^\alpha$ (see Eq. (6.42)), this can be written

$$V^{\alpha}_{;\beta;\sigma} - V^{\alpha}_{;\sigma;\beta} = R^{\alpha}_{\rho\beta\sigma} V^{\rho}. \tag{6.43}$$

Note the commutators of ordinary derivatives vanish as would the commutators of covariant derivatives at a point where the Riemann tensor vanishes.

The Riemann tensor has $4^4 = 256$ components; however, not all are independent for the tensor possesses a large number of symmetries best expressed in terms of the completely covariant form obtained from the above by "lowering" the ν index :

$$R_{\sigma\beta\alpha\rho} \equiv g_{\sigma\nu} R^{\nu}_{\beta\alpha\rho}.$$

The following equalities are most easily derived in a local inertial frame (Exercise 3):

$$R_{\sigma\beta\alpha\rho} = - R_{\beta\sigma\alpha\rho} = - R_{\sigma\beta\rho\alpha} = R_{\alpha\rho\sigma\beta}$$
$$R_{\sigma\beta\alpha\rho} + R_{\sigma\rho\beta\alpha} + R_{\sigma\alpha\rho\beta} = 0. \tag{6.44}$$

From these it follows that

$$R^{\sigma}_{\sigma\alpha\rho} \equiv 0. \tag{6.45}$$

6.6   Bianchi Identity and Ricci and Einstein Tensors

Again, the following important identity, called the *Bianchi identity*, is most easily shown in a local inertial coordinate system:

$$R_{\sigma\beta\alpha\rho;\eta} + R_{\sigma\beta\eta\alpha;\rho} + R_{\sigma\beta\rho\eta;\alpha} = 0. \tag{6.46}$$

An important tensor, called the *Ricci tensor*, is formed by a contraction on the Riemann tensor:

$$R_{\beta\rho} \equiv R^{\nu}_{\beta\nu\rho} = R_{\rho\beta}. \tag{6.47}$$

Because of the symmetries of $R_{\sigma\beta\alpha\rho}$ this is the only independent contraction on the Riemann tensor. Note that the contraction used for the definition of the Ricci tensor is a contraction of the contravariant index of the Riemann tensor with its middle covariant index.

Similarly, the *Ricci scalar* is defined as

$$R \equiv g^{\beta\rho} R_{\beta\rho}. \tag{6.48}$$

In order to see the implication of the Bianchi identity on the Ricci tensor, we do a "Ricci"-like contraction on this identity;

$$g^{\sigma\alpha}[R_{\sigma\beta\alpha\rho;\,\eta} + R_{\sigma\beta\eta\alpha;\,\rho} + R_{\sigma\beta\rho\eta;\,\alpha}] = 0.$$

As noted before, $g^{\alpha\beta}_{;\,\rho} = 0$ and $g_{\alpha\beta;\,\rho} = 0$. Thus $g^{\alpha\beta}$ and $g_{\alpha\beta}$ can be taken in and out of covariant differentiation, and, we have

$$R_{\beta\rho;\,\eta} - R_{\beta\eta;\,\rho} + R^{\alpha}_{\beta\rho\eta;\,\alpha} = 0. \tag{6.49}$$

An even more useful equation is obtained by a second contraction,

$$g^{\beta\rho}[R_{\beta\rho;\,\eta} - R_{\beta\eta;\,\rho} + R^{\alpha}_{\beta\rho\eta;\,\alpha}] = 0,$$

or

$$R_{;\,\eta} - R^{\rho}_{\eta;\,\rho} - R^{\alpha}_{\eta;\,\alpha} = 0. \tag{6.50}$$

This equation can be written in the form

$$(2R^{\alpha}_{\eta} - \delta^{\alpha}_{\eta} R)_{;\,\alpha} = 0. \tag{6.51}$$

The completely contravariant form of this equation is

$$G^{\alpha\rho}_{;\,\alpha} = 0, \tag{6.52}$$

where

$$G^{\alpha\rho} \equiv R^{\alpha\rho} - \frac{1}{2} g^{\alpha\rho} R. \tag{6.53}$$

$G^{\alpha\rho}$ is called the *Einstein tensor*. We will see that it is of fundamental importance in the Einstein field equations for general relativity, as is the identity Eq. (6.52), also referred to as the *Bianchi identity*.

Listed below are some differential geometry definitions and results that will be used repeatedly in discussions that follow:

Christoffel symbols
$$\Gamma^{\rho}_{\sigma\mu} = \frac{1}{2} g^{\nu\rho} \left[\frac{\partial g_{\mu\nu}}{\partial x^{\sigma}} + \frac{\partial g_{\sigma\nu}}{\partial x^{\mu}} - \frac{\partial g_{\mu\sigma}}{\partial x^{\nu}} \right]$$

Geodesic
$$\frac{d^2 x^{\nu}}{d\tau^2} + \Gamma^{\nu}_{\rho\beta}(x) \frac{dx^{\beta}}{d\tau} \frac{dx^{\rho}}{d\tau} = 0$$

Directional covariant derivative
$$\frac{DA^{\nu}}{ds} = \frac{dA^{\nu}}{ds} + \Gamma^{\nu}_{\rho\beta}(x) A^{\beta} \frac{dx^{\rho}}{ds}$$

Covariant derivative $\qquad A^{\nu}_{;\,\rho} \equiv \dfrac{\partial A^{\nu}}{\partial x^{\rho}} + \Gamma^{\nu}_{\rho\beta}(x)\, A^{\beta}$

Riemann curvature tensor $\quad R^{\nu}_{\beta\alpha\rho} = \Gamma^{\nu}_{\beta\alpha,\,\rho} - \Gamma^{\nu}_{\beta\rho,\,\alpha} + \Gamma^{\nu}_{\rho\eta}\,\Gamma^{\eta}_{\beta\alpha} - \Gamma^{\nu}_{\alpha\eta}\,\Gamma^{\eta}_{\beta\rho},$

Commutator of covariant $\quad V^{\alpha}_{;\,\beta;\,\sigma} - V^{\alpha}_{;\,\sigma;\,\beta} = R^{\alpha}_{\rho\beta\sigma}\, V^{\rho}$
derivative

Ricci tensor $\qquad\qquad R_{\beta\rho} \equiv R^{\nu}_{\beta\nu\rho} = R_{\rho\beta}$

Ricci scalar $\qquad\qquad R \equiv g^{\beta\rho}\, R_{\beta\rho}$

Einstein tensor $\qquad\qquad G^{\alpha\rho} \equiv R^{\alpha\rho} - \dfrac{1}{2}\, g^{\alpha\rho}\, R$

6.7   The Einstein Field Equations

In Chapter 5 we saw that the equivalence principle seems to imply that the metric is changed in the presence of matter. When restriction was made to low relative velocities and weak static fields, we argued that

$$g_{00} \approx 1 + 2\Phi. \tag{6.54}$$

Here Φ, the Newtonian gravitational potential, satisfies the equation

$$\nabla^{2}\Phi = 4\pi\tilde{G}\rho, \tag{6.55}$$

where ρ is the mass density. Thus, in a relativistic theory of gravity the equation

$$\nabla^{2} g_{00} = 8\pi\tilde{G}\rho \tag{6.56}$$

must obtain in some low-speed and weak-field approximation. Since ρ is the T^{00} component of the energy-momentum tensor in the rest frame of a fluid, we might expect that Eq. (6.56) should be replaced by a generally covariant equation with the energy-momentum tensor as the source of a tensor $O^{\alpha\beta}$ of contravariant rank 2:

$$O^{\alpha\beta}(x) = k T^{\alpha\beta}(x).$$

$O^{\alpha\beta}$ should be expressible in terms of the metric and should involve up to second-order derivatives of the metric. Second-rank tensors that might be used are $R^{\alpha\beta}, g^{\alpha\beta} R$, and $g^{\alpha\beta}$. Thus, we might expect as possibilities

$$O^{\alpha\beta} = R^{\alpha\beta} + cg^{\alpha\beta} R + \lambda g^{\alpha\beta},$$

where c and λ are constants. But the local conservation of energy momentum, Eq. (4.52), along with the principle of general covariance demands that

$$T^{\alpha\beta}_{;\beta} = 0,$$

and thus

$$O^{\alpha\beta}_{;\beta} = 0$$

must be generally true. We see, by use of Eq. (6.52), that this equation is satisfied if $c = -1/2$ and we are, plausibly, led to the following field equations for the metric:

$$G^{\alpha\beta}(x) + \lambda g^{\alpha\beta}(x) = kT^{\alpha\beta}(x), \tag{6.57}$$

with the constants λ and k to be determined. A restriction on these constants arises if Eq. (6.56) is to be satisfied in a weak-field, low-velocity limit. In the low-velocity limit $|T^{ij}| << |T^{00}|$. (As an example, for a perfect fluid, see Eq. (4.46).) But this implies that

$$G^{ij}(x) + \lambda g^{ij}(x) \approx 0, \tag{6.58}$$

or

$$R^{ij} \approx \frac{1}{2} g^{ij} R - \lambda g^{ij}(x). \tag{6.59}$$

We assume the meaning of the weak-field limit to be the existence of coordinates such that

$$g_{\alpha\beta} \approx \eta_{\alpha\beta} + \delta g_{\alpha\beta}, \ |\delta g_{\alpha\beta}| << 1. \tag{6.60}$$

By use of such a metric and Eq. (6.59), we have

$$R = n_{\alpha\beta} R^{\alpha\beta} = R^{00} - R^{ii} \approx R^{00} + \frac{3}{2} R + \lambda g^{ii}(x), \tag{6.61}$$

or

$$R \approx -2R^{00} - 2\lambda g^{ii}. \tag{6.62}$$

With this, the $\{0\,0\}$ component of Eq. (6.57) becomes

$$2R^{00} + \lambda [g^{00} + g^{00} g^{ii}] \approx kT^{00}, \tag{6.63}$$

where we have used $g^{00}R^{00} \approx R^{00}$. We need to know what R^{00} is in the weak-field approximation. We start with the Riemann tensor:

$$R^{\sigma}_{\beta\alpha\rho} \approx \Gamma^{\sigma}_{\beta\rho,\,\alpha} - \Gamma^{\sigma}_{\beta\alpha,\,\rho}. \tag{6.64}$$

Terms quadratic in Γs and thus quadratic in $\delta g_{\alpha\beta}$ have been dropped. Using

$$R_{\nu\beta\alpha\rho} \approx n_{\nu\sigma} R^{\sigma}_{\beta\alpha\rho}$$

and Eq. (6.26), one obtains

$$R_{\nu\beta\alpha\rho} \approx -\frac{1}{2}[g_{\rho\nu,\,\beta\alpha} - g_{\rho\beta,\,\nu\alpha} - g_{\alpha\nu,\,\beta\rho} + g_{\alpha\beta,\,\nu\rho}]. \tag{6.65}$$

Because we are dealing with static fields, all time derivatives vanish. We want an expression for R^{00}. Within our approximation, we have

$$R^{00} \approx R_{0000} - R_{i0i0} \approx -R_{i0i0} \approx -\frac{1}{2}g_{00,\,ii} = -\frac{1}{2}\nabla^2 g_{00}.$$

Using this in Eq. (6.63), we obtain

$$-\nabla^2 g_{00} + \lambda[g^{00} + g^{00}g^{ii}] \approx kT^{00}. \tag{6.66}$$

We see that Eq. (6.56) results if $k = -8\pi\tilde{G}$ and $\lambda = 0$. The constant λ, referred to as the *cosmological constant*, was not included at first in Einstein's field equations. He added it later so that a static cosmological solution to the equations would exist. He discarded it still later. Observations of gravitating systems imply λ is quite small. (See the next section.) Except for some exercises and in Chapter 9, where cosmology is discussed, we assume $\lambda = 0$. With λ included, Einstein's field equations become

$$G^{\alpha\beta}(x) \equiv R^{\alpha\beta} - \frac{1}{2}g^{\alpha\beta}R = -8\pi\tilde{G}T^{\alpha\beta}(x) - \lambda g^{\alpha\beta}. \tag{6.67}$$

With λ set equal to zero, the field equations are

$$G^{\alpha\beta}(x) \equiv R^{\alpha\beta} - \frac{1}{2}g^{\alpha\beta}R = -8\pi\tilde{G}T^{\alpha\beta}(x). \tag{6.68}$$

We find from this that

$$R - 2R = -8\pi\tilde{G}T^{\alpha}_{\alpha} \equiv -8\pi\tilde{G}T, \tag{6.69}$$

and thus Eq. (6.68) can be written

$$R^{\alpha\beta} = -8\pi\tilde{G}\left(T^{\alpha\beta}(x) - \frac{1}{2}g^{\alpha\beta}T\right). \tag{6.70}$$

Note that in "vacuum," $T^{\alpha\beta}(x) = 0$ and the field equations reduce to

$$R^{\alpha\beta} = 0. \tag{6.71}$$

We will obtain a nontrivial solution to this equation in the next chapter.

6.8   The Cosmological Constant

Eq. (6.67) can be rewritten as

$$G^{\alpha\beta}(x) \equiv R^{\alpha\beta} - \frac{1}{2}g^{\alpha\beta}R = -8\pi\tilde{G}(T^{\alpha\beta}(x) + \frac{\lambda}{8\pi\tilde{G}}g^{\alpha\beta}). \tag{6.72}$$

The cosmological term behaves as an effective energy-momentum density $T^{\alpha\beta}_{\lambda}$:

$$T^{\alpha\beta}_{\lambda} = \frac{\lambda}{8\pi\tilde{G}}g^{\alpha\beta}.$$

It is present even where $T^{\alpha\beta} = 0$ and is sometimes referred to as a vacuum energy-momentum density.

With $|\lambda|$ sufficiently small, Eq. (6.66) would remain a good approximation to Eq. (6.56) and thus to Eq. (6.55) for the Newtonian gravitational potential. We can ask how small λ must be so that it doesn't have a significant effect on the dynamics of gravitating systems that are known to be very well described by Newton's gravitational theory. For λ small, the $g^{\alpha\beta}$ in the second term of Eq. (6.66) can be replaced with $n^{\alpha\beta}$, and the equation can be then written as

$$\nabla^2 g_{00} \approx 8\pi\tilde{G}T^{00} - 2\lambda, \tag{6.73}$$

or, in terms of the gravitational potential, Φ,

$$\nabla^2 \Phi = -\lambda + 4\pi \tilde{G} \rho. \tag{6.74}$$

We see that λ contributes an effective mass density, ρ_λ, given by

$$\rho_\lambda = -\frac{\lambda}{4\pi \tilde{G}}.$$

For a spherically symmetric system, this term gives a contribution to the gravitational potential,

$$\Phi_\lambda(r) = -\frac{\lambda r^2}{6}. \tag{6.75}$$

Note we have set $\Phi_\lambda(0) = 0$ and $\Phi_\lambda \to \infty$ as $r \to \infty$. Clearly, irrespective of how small λ is, the approximation $g^{\alpha\beta} \approx n^{\alpha\beta}$ breaks down for r large.

The potential $\Phi_\lambda(r)$ gives a radial force per unit mass of $\lambda r/3$. When a system is well described by a Newtonian potential $\Phi_N(r)$, which gives a radial force per unit mass of $-d\Phi_N(r)/dr$, then

$$\frac{|\lambda|\, r}{3} \ll \left| \frac{d\Phi_N(r)}{dr} \right| \tag{6.76}$$

must be true. As an example, for a gravitational system such as our solar system, whose dynamics is determined by a large mass M, with a largest observed orbit radius of r_L, this inequality becomes

$$|\lambda| \ll \frac{\tilde{G}M}{r_L^3}. \tag{6.77}$$

With M the mass of the sun and r_L the radius of the orbit of Jupiter, we have $|\lambda| < 10^{-35}\, m^{-2}$. The effect of the λ term is larger the larger the size of the system considered. That the Newtonian gravitational force law seems to be valid for systems of galaxies results in a more stringent bound, $|\lambda| < 10^{-46}\, m^{-2}$. But importantly, a much smaller $|\lambda|$ can have a significant effect on systems of cosmological size.

6.9 Energy-Momentum Tensor of a Perfect Fluid in General Relativity

The properties of energy-momentum tensor as the source of the metric are of course important in the applications of general relativity. One of the most important is the local conservation law, which in special relativity, is expressed by Eq. (4.52)

$$\partial_\beta T^{\alpha\beta} = 0. \tag{6.78}$$

The general covariant form of this equation is

$$T^{\alpha\beta}_{;\beta} = 0. \tag{6.79}$$

That is, the ordinary derivative in Eq. (6.78) is replaced by the covariant derivative. Eq. (6.79) reduces to Eq. (6.78) in a local inertial frame, as it must to be consistent with the principle of general covariance.

In general relativity, as in special relativity, an important example of an energy-momentum tensor is that of a perfect fluid, which in special relativity has the form

$$T^{\alpha\beta} = -p(x)\,n^{\alpha\beta} + (\rho(x) + p(x))\,U^\alpha U^\beta. \tag{6.80}$$

In general relativity the energy-momentum tensor of a perfect fluid must then have the form

$$T^{\alpha\beta} = -p(x)\,g^{\alpha\beta} + (\rho(x) + p(x))\,U^\alpha U^\beta, \tag{6.81}$$

which is a rank 2 general covariant tensor and which reduces to Eq. (6.80) in a local inertial frame and is thus consistent with the principle of general covariance. Recall that $p(x)$ and $\rho(x)$ are defined to be the pressure and the energy density (at an event point x) as measured by an observer in a locally inertial frame that is at rest with respect to the fluid, and are thus scalar fields under general coordinate transformations.

For such a perfect fluid let us compute $T^{\alpha\beta}_{;\beta}$ to see what is implied by energy-momentum conservation. Recall that $g^{\alpha\beta}_{;\beta} = 0$ and, since $p(x)$ and $\rho(x)$ are scalar fields, $p(x)_{;\beta} = p(x)_{,\beta}$ and $\rho(x)_{;\beta} = \rho(x)_{,\beta}$. Thus, we have

$$T^{\alpha\beta}_{;\beta} = ((p+\rho)\,U^\alpha U^\beta)_{;\beta} - p_{,\beta}\,g^{\alpha\beta}$$

$$= (p+\rho)_{,\beta}\,U^\alpha U^\beta + (p+\rho\big[(U^\alpha U^\beta)_{,\beta} + \Gamma^\alpha_{\sigma\beta}U^\sigma U^\beta + \Gamma^\beta_{\sigma\beta}U^\sigma U^\alpha\big]$$

$$- p_{,\beta}\,g^{\alpha\beta} = 0. \tag{6.82}$$

In a local inertial frame, this becomes

$$T^{\alpha\beta}_{;\beta} = (p+\rho)_{,\beta}\,U^\alpha U^\beta + (p+\rho)(U^\alpha U^\beta)_{,\beta} - p_{,\beta}\,g^{\alpha\beta} = 0, \tag{6.83}$$

which in turn, in such a frame moving with the fluid, reduces to Eqs. (4.63) and (4.64).

6.10   Exercises

1. Develop your own *Mathematica* (or *Maple*) program that, given a metric, will calculate (a) the Christoffel symbols, $\Gamma^\nu_{\rho\beta}$; (b) the curvature tensor, $R^\nu_{\beta\alpha\rho}$; (c) the Ricci tensor, $R^\nu_{\beta\nu\rho}$; and (d) the Ricci scalar, R.

2. For a closed path as indicated in Figure 6.4, evaluate $\int_0^1 [x^\alpha - x_0^\alpha] \dfrac{dx^\beta}{ds}\, ds$.

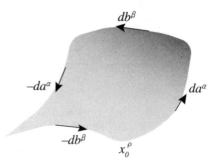

Figure 6.4. Closed path.

3. (a) Derive the equalities of the Riemann tensor, Eqs. (6.44) and (6.45). (b) Derive the Bianchi identity, Eq. (6.46).

4. Let the metric on the surface of a sphere be given by

$$ds^2 = d\theta^2 + \sin^2\theta\, d\phi^2 = (dx^1)^2 + \sin^2 x^1 (dx^2)^2.$$

(a) What are the Christoffel symbols? (b) What are the nonvanishing components of the curvature tensor? (c) What is the Ricci tensor? (d) What is the Ricci scalar? (e) What vector results if one parallel transports a vector that starts with only a θ component, $A^i = A\delta^{i1}$, around a close path with $da^i = da\delta^{i1}$ and $db^i = db\delta^{i2}$ in the notation of Exercise 2? (f) Prove that the path defined by $\phi = $ constant is a geodesic.

5. (a) Calculate the curvature tensors and the Ricci scalars for the metrics

$$ds^2 = y^n\, dx^2 + x^n\, dy^2$$

for $n = 1, 2, 3, 4$. Are any of these metrics the metric of two-dimensional Euclidean space? (b) Calculate the curvature tensors and the Ricci scalars for the metrics

$$ds^2 = x^n \, dx^2 + y^n \, dy^2$$

for $n = 1, 2, 3, 4$. Are any of these metrics the metric of two-dimensional Euclidean space? Why might you have anticipated the result?

6. In the low-speed and weak-field approximation (Eq. (6.60)), show that a freely falling particle of mass m satisfies the expected equation

$$\mathbf{a} = -\nabla g_{00}/2.$$

Note in this approximation $dx^0/d\tau \gg dx^i/d\tau$. This result is valid even if $\lambda \neq 0$, but is, of course, considered small.

Chapter 7

Static Spherical Metrics and Their Applications

7.1 Introduction

In the study of most theories, highly symmetric solutions play a defining role. Compared to more general solutions, they are easier to obtain, and their implications for physical phenomena are more apparent. They give insights into a theory that, quite often, have general validity. And luckily nature is kind in providing examples that are quite close to being symmetric. In this chapter we will study spherically symmetric solutions to Einstein's field equations.

Einstein suggested three tests of his theory of general relativity:

1. The gravitational red shift of light.
2. The deflection of light by the sun.
3. The precession of the perihelia of the orbits of planets, in particular that of Mercury, the innermost planet.

These processes take place in empty space and for gravitational fields that are almost static and, to a good approximation, spherically symmetric. For this reason, as well as for the intrinsic interest of an exact nontrivial solution to the field equations, we obtain a spherically symmetric static solution of the vacuum field equations, the *Schwarzschild metric*. Then, though we have already derived the gravitational redshift for weak fields using the equivalence principle, we will derive the magnitude of the effect for a static metric. We follow this with a discussion of geodesic motion for massive and massless particles in a Schwarzschild metric and apply the results to (2) and (3) above.

The dynamics of a gyroscope in general relativity is then considered and specialized to a study of a gyroscope in orbit (i.e., one that is freely falling), a generalization of the *Thomas precession* that occurs in flat space. Measurement of this precession may prove to be one of the most sensitive tests of general relativity.

After studying the processes exterior to a massive body, we will derive the equations governing the dynamics of a fluid comprising the interior of such a body. For a particularly simple (and unphysical) fluid, we show that, in contrast to the Newtonian equations, the relativistic equations limit the mass-radius ratio of a static massive body.

Consideration of the geometry of the Schwarzschild metric in the extreme gives rise to the concepts of *event horizons* and *black holes*, described in the last section of this chapter.

7.2 The Static Spherical Metric

The meaning of a static spherical metric can be characterized in a way independent of the coordinate system used (see Chapter 8). Here we take the definition of such a metric to be one for which a set of coordinates $\{x^0, x^i\}$ exists for which the invariant interval is expressible in terms of rotational scalars $dx^0, x^i x^i = r, x^i dx^i, dx^i dx^i$, and which is x^0 independent. Thus, the invariant interval can be expressed in the form[1]

$$d\tau^2 = D(r)(dx^0)^2 - 2E(r)x^i dx^i dx^0 - F(r)(x^i dx^i)^2 - H(r)dx^i dx^i), \quad (7.1)$$

or in terms of the spherical coordinates, r, θ and ϕ,

$$d\tau^2 = D(r)(dx^0)^2 - 2E(r)rdrdx^0$$
$$- F(r)r^2 dr^2 - H(r)(dr^2 + r^2 d\theta^2 + r^2 \sin^2\theta d\theta^2).$$

One can transform to a new time variable so that a term like $2E(r)rdrdx^0$ does not appear in the metric. With the new time x'^0 defined by

$$x^0 = G(x'^0, r),$$

the condition $g_{0'r} = 0$ requires that

$$\frac{\partial G}{\partial x'^0}\left(D(r)\frac{\partial G}{\partial r} - rE(r)\right) = 0.$$

---

[1] This characterization of a static spherical metric follows Weinberg (1973).

This equation is satisfied if we choose $x^0 = x'^0 + I(r)$, with $I(r)$ given by

$$\frac{dI(r)}{dr} = \frac{rE(r)}{D(r)}.$$

With this change of the time variable, the invariant interval becomes

$$d\tau^2 = D(r)(dx'^0)^2 - K(r) dr^2 - H(r)(dr^2 + r^2 d\theta^2 + r^2 \sin^2\theta d\theta^2),$$

where

$$K(r) = r^2 \left(F(r) - \frac{E(r)^2}{D(r)} \right).$$

We can further reduce the number of functions of r in the metric to two, if we define a new "radial" variable

$$r'^2 = H(r) r^2.$$

The invariant interval then becomes

$$d\tau^2 = B(r')(dx'^0)^2 - A(r') dr'^2 - r'^2 d\theta^2 - r'^2 \sin^2\theta d\phi^2,$$

with

$$B(r') \equiv D(r)$$

$$A(r') \equiv \left[\frac{K(r)}{H(r)} + 1 \right] \left[\frac{H'(r)r}{2H(r)} + 1 \right]^{-2}.$$

Finally, we drop the primes and write the invariant interval as

$$d\tau^2 = B(r)(dx^0)^2 - A(r) dr^2 - r^2 d\theta^2 - r^2 \sin^2\theta d\phi^2. \tag{7.2}$$

Computing the Ricci tensor (say, by use of *Mathematica*), we find

$$R_{00} = -\frac{B'(r)}{rA(r)} + \frac{1}{4}\left(\frac{B'(r)}{A(r)}\right)\left(\frac{A'(r)}{A(r)} + \frac{B'(r)}{B(r)}\right) - \frac{B''(r)}{2A(r)} \tag{7.3}$$

$$R_{rr} = -\frac{A'(r)}{rA(r)} - \frac{1}{4}\left(\frac{B'(r)}{B(r)}\right)\left(\frac{A'(r)}{A(r)} + \frac{B'(r)}{B(r)}\right) + \frac{B''(r)}{2B(r)} \tag{7.4}$$

$$R_{\theta\theta} = -1 + \frac{r}{2A(r)}\left(-\frac{A'(r)}{A(r)} + \frac{B'(r)}{B(r)}\right) + \frac{1}{A(r)} \tag{7.5}$$

$$R_{\phi\phi} = \sin^2(\phi) R_{\theta\theta} \tag{7.6}$$

$$R_{\mu\nu} = 0; \ \mu \neq \nu. \tag{7.7}$$

7.3 The Schwarzschild Solution

We are now in a position to solve for the metric *outside* of a static spherically symmetric body. In this region the metric satisfies Einstein's vacuum field equation, Eq. (6.71). We will impose the condition that very far from the mass the metric becomes *flat*; that is, it is *asymptotically flat*. Note that this condition applied to the metric of the form Eq. (7.2) gives

$$\lim_{r \to \infty} A(r) = \lim_{r \to \infty} B(r) = 1. \tag{7.8}$$

In fact, more general considerations lead to *Birkhoff's theorem* (Birkhoff 1923) that states that the spherically symmetric asymptotically flat metric is unique. Thus, the static condition can be relaxed and the solution will apply outside of a spherically symmetric nonstatic mass. From Eqs. (7.3) to (7.6), the field equations Eq. (6.71) for the metric of Eq. (7.2) become

$$R_{00} = 0 = -\frac{B'(r)}{rA(r)} + \frac{1}{4}\left(\frac{B'(r)}{A(r)}\right)\left(\frac{A'(r)}{A(r)} + \frac{B'(r)}{B(r)}\right) - \frac{B''(r)}{2A(r)} \tag{7.9}$$

$$R_{rr} = 0 = -\frac{A'(r)}{rA(r)} - \frac{1}{4}\left(\frac{B'(r)}{B(r)}\right)\left(\frac{A'(r)}{A(r)} + \frac{B'(r)}{B(r)}\right) + \frac{B''(r)}{2B(r)} \tag{7.10}$$

$$R_{\theta\theta} = 0 = -1 + \frac{r}{2A(r)}\left(-\frac{A'(r)}{A(r)} + \frac{B'(r)}{B(r)}\right) + \frac{1}{A(r)}. \tag{7.11}$$

$$R_{\phi\phi} = 0 = \sin^2(\phi) R_{\theta\theta}. \tag{7.12}$$

We see that there are three independent equations to be satisfied. From Eqs. (7.9) and (7.10), we have

$$\frac{R_{00}}{B(r)} + \frac{R_{rr}}{A(r)} = -\frac{1}{rA(r)}\left(\frac{B'(r)}{B(r)} + \frac{A'(r)}{A(r)}\right) = 0. \tag{7.13}$$

This, with the asymptotically flat condition, Eq. (7.8), implies

$$A(r) = \frac{1}{B(r)}. \tag{7.14}$$

Using this result in Eq. (7.11), we find

$$-1 + (rB(r))' = 0, \tag{7.15}$$

which has a solution

$$B(r) = 1 + \frac{c}{r}. \qquad (7.16)$$

Here c is some constant. Asymptotically, since the gravitational field is weak, Eq. (6.54) applies. This justifies writing

$$g_{00} = B(r) \approx 1 + 2\Phi = 1 - \frac{2\tilde{G}M}{r}, \text{ for } r \to \infty \qquad (7.17)$$

and calling M the total gravitational mass. This parameter is the total mass of the solution in the sense that distant orbits are Newtonian orbits of a spherical mass M centered at $r = 0$. We see that c of Eq. (7.16) is $-2\tilde{G}M$ and we have for the invariant interval

$$d\tau^2 = \left(1 - \frac{2\tilde{G}M}{r}\right)(dx^0)^2 - \left(1 - \frac{2\tilde{G}M}{r}\right)^{-1} dr^2 - r^2\,d\theta^2 - r^2\sin^2\theta d\phi^2,$$

$$(7.18)$$

a solution to the vacuum field equations found by Schwarzschild. The nonvanishing Christoffel symbols (see Exercise 2) for this metric are

$$\Gamma^r_{0\,0} = \frac{\tilde{G}M(-2\tilde{G}M + r)}{r^3}$$

$$\Gamma^r_{r\,r} = -\Gamma^0_{r\,0} = -\Gamma^0_{0\,r} = \frac{\tilde{G}M}{2\tilde{G}Mr - r^2}$$

$$\Gamma^r_{\theta\,\theta} = \Gamma^r_{\phi\,\phi}/\sin^2\theta = 2\tilde{G}M - r$$

$$\Gamma^\theta_{\phi\,\phi} = -\cos\theta\sin\theta$$

$$\Gamma^\theta_{r\,\theta} = \Gamma^\theta_{\theta\,r} = \Gamma^\phi_{r\,\phi} = \Gamma^\phi_{\phi\,r} = r^{-1}$$

$$\Gamma^\phi_{\phi\,\theta} = \Gamma^\phi_{\theta\,\phi} = \cot\theta.$$

7.4   Gravitational Redshift

We return to the calculation of the change in frequency of a light wave as it moves radially in the presence of a massive body whose metric is stationary—for instance, a Schwarzschild metric. Specifically, we ask, What is the relation between the frequency ν_1 measured by an observer at "rest" at r_1 and the frequency ν_2 measured by an observer at "rest" at r_2?

The proper time period of the wave measured by the observer fixed at r_1 is

$$dt_1 = g_{00}(r_1)^{1/2} dx^0,$$

where dx^0 is the change in coordinate time. The proper time period of the wave measured by the observer fixed at r_2 is

$$dt_2 = g_{00}(r_2)^{1/2} dx^0,$$

where dx^0 is the same change in coordinate time as at r_1. This is true because, as noted in Section 5.3, this is an essentially static process because the metric is static. The coordinate time period is the time it takes one wavelength to pass the observer's position and must be the same for both observers. Thus, we can write

$$\frac{dt_1}{dt_2} = \frac{g_{00}(r_1)^{1/2}}{g_{00}(r_2)^{1/2}},$$

from which follows

$$\frac{\nu_2}{\nu_1} = \frac{g_{00}(r_1)^{1/2}}{g_{00}(r_2)^{1/2}}. \tag{7.19}$$

How does this compare to the shift Einstein obtained in his 1911 paper that we discussed in Chapter 5? That shift is given by Eq. (5.2), which written in terms of frequencies takes the form

$$\frac{\nu_2}{\nu_1} = 1 - (\Phi(r_2) - \Phi(r_1)).$$

If we use the weak-field approximation $g_{00} = 1 + 2\Phi$ in Eq. (7.19), we have

$$\frac{\nu_2}{\nu_1} = \frac{g_{00}(r_1)^{1/2}}{g_{00}(r_2)^{1/2}} \approx \frac{(1 + 2\Phi(r_1))^{1/2}}{(1 + 2\Phi(r_2))^{1/2}} \approx 1 - (\Phi(r_2) - \Phi(r_1)),$$

the same result Einstein obtained in his 1911 paper.

7.5 Conserved Quantities

Next we study the geodesic motion of both massive particles and massless particles in the highly symmetric Schwarzschild metric. Realizing that symmetries give rise to conserved quantities, we would expect to be able to find entities associated with the symmetries of the metric that are conserved as the particle moves along a geodesic.

Such conserved quantities will of course be useful in solving the geodesic motion.

First, consider the motion for a massive particle. We subsequently see how our results need to be modified for a zero-mass particle. The geodesic equation can be written (Eq. (6.31))

$$\frac{dx^\rho}{d\tau} U^\alpha_{;\rho} = 0,$$ (7.20)

where $U^\rho = dx^\rho / d\tau$, the generalized "four-velocity". In terms of the covariant generalized "four-momentum" $p_\sigma = mU_\sigma$ (recall that the metric can be taken inside the covariant derivative) this can be expressed as

$$mU^\rho p_{\sigma;\rho} = 0$$ (7.21)

or as

$$p^\rho p_{\sigma,\rho} - \Gamma^\alpha_{\sigma\rho} p^\rho p_\alpha = 0.$$

or as

$$m\frac{dp_\sigma}{d\tau} = \Gamma^\alpha_{\sigma\rho} p^\rho p_\alpha$$
$$= \frac{1}{2}(g_{\zeta\sigma,\rho} + g_{\zeta\rho,\sigma} - g_{\sigma\rho,\zeta})p^\zeta p^\rho = \frac{1}{2} g_{\zeta\rho,\sigma} p^\zeta p^\rho.$$ (7.22)

We see that the geodesic equation can always be written as

$$m\frac{dp_\sigma}{d\tau} = \frac{1}{2} g_{\zeta\rho,\sigma} p^\zeta p^\rho.$$ (7.23)

From this equation follows a useful result:

If the metric $g_{\alpha\beta}$ is dependent of x^σ for some fixed σ then p_σ is constant along any geodesic.

Of course, the statement that the metric is independent of x^σ is coordinate dependent; one would expect that the identification of a conserved quantity associated with a symmetry of the metric could be made in a coordinate-independent way once the characterization of the symmetry of the metric is made in a coordinate-independent manner. This will be done in some detail in Chapter 8.

How are these considerations modified for a zero-mass particle? Recall that in the geodesic equation, Eq. (7.23), if we parameterize the geodesic path by an affine parameter $s = c\tau$ for any constant c, then the equation for the geodesic in terms of s is identical to that of τ. We could, for instance, use $s = \tau/m$ and for such we see $p^\alpha = dx^\alpha/ds$, and Eq. (7.23) becomes

$$\frac{dp_\sigma}{ds} = \frac{1}{2} g_{\zeta\rho,\sigma} p^\zeta p^\rho. \qquad (7.24)$$

The generalization of these equations to zero-mass particles is clear. For a particle of nonzero mass we can use τ to parameterize the geodesic. For a zero-mass particle, we can take $s = \lim_{m \to 0}(\tau/m)$. In either case, $p^\alpha p_\alpha = m^2$.

7.6 Geodesic Motion for a Schwarzschild Metric

The Schwarzschild metric is time independent and spherically symmetric. We would expect that, since it is spherically symmetric, geodesic motion is confined to a "plane", as in the Newtonian case. We can easily see this is so. Consider the geodesic equation for p_θ:

$$\frac{dp_\theta}{ds} = \frac{1}{2} g_{\zeta\rho,\theta} p^\zeta p^\rho = \sin\theta \cos\theta p^\phi p^\phi.$$

Since the metric is spherically symmetric for any point on the geodesic, we can pick the spherical coordinates such that $p_\theta = 0$ and $\theta = \pi/2$; the particle is moving along the "equator" at that instant. But this equation shows that, at that instant, $dp_\theta/ds = 0$. Thus, it continues to move along the equator, or in a plane. Without loss of generality, we can assume the particle moves with $\theta = \pi/2$.

Since the metric is independent of x^0 and ϕ, p_0 and p_ϕ are constant. For a massive particle we define

$$p_0 = mE, \quad p_\phi = -mL.$$

(Here E is a dimensionless constant, whereas the constant L has the dimension of length.) For a zero-mass particle we define

$$p_0 = E, \quad p_\phi = -L.$$

(Here E has the dimension of mass, whereas L has the dimension of length × mass.) Thus, for a massive particle,

$$p^0 = g^{00} p_0 = \left(1 - \frac{2M\tilde{G}}{r}\right)^{-1} mE \qquad (7.25)$$

$$p^r = m \frac{dr}{d\tau} \qquad (7.26)$$

$$p^\phi = g^{\phi\phi} p_\phi = r^{-2} mL = m \frac{d\phi}{d\tau}, \tag{7.27}$$

whereas for a zero-mass particle,

$$p^0 = g^{00} p_0 = \left(1 - \frac{2M\tilde{G}}{r}\right)^{-1} E \tag{7.28}$$

$$p^r = \frac{dr}{ds} \tag{7.29}$$

$$p^\phi = g^{\phi\phi} p_\phi = r^{-2} L = \frac{d\phi}{ds}. \tag{7.30}$$

By use of Eqs. (7.25) — (7.27) , $p^\alpha p_\alpha = m^2$ becomes an equation for $dr/d\tau$:

$$\left(1 - \frac{2M\tilde{G}}{r}\right)^{-1} m^2 E^2 - \left(1 - \frac{2M\tilde{G}}{r}\right)^{-1} m^2 \left(\frac{dr}{d\tau}\right)^2 - r^{-2} m^2 L^2 = m^2. \tag{7.31}$$

Thus the "radial" equation for a massive particle is

$$\left(\frac{dr}{d\tau}\right)^2 = E^2 - \left(1 + \frac{L^2}{r^2}\right)\left(1 - \frac{2\tilde{G}M}{r}\right). \tag{7.32}$$

Similarly, the radial equation for a zero-mass particle is

$$\left(\frac{dr}{ds}\right)^2 = E^2 - \frac{L^2}{r^2}\left(1 - \frac{2\tilde{G}M}{r}\right). \tag{7.33}$$

We can obtain the orbit equation, an equation for $dr/d\phi$, by dividing $dr/d\tau$ (dr/ds) by $d\phi/d\tau$ $(d\phi/ds)$ for the massive particle (for the zero-mass particle), with the result

$$\left(\frac{dr}{d\phi}\right)^2 = \frac{E^2 - \left(1 + \frac{L^2}{r^2}\right)\left(1 - \frac{2\tilde{G}M}{r}\right)}{L^2/r^4} \tag{7.34}$$

for the massive particle. With the substitution

$$u = \frac{1}{r}, \tag{7.35}$$

this orbit equation becomes

$$\left(\frac{du}{d\phi}\right)^2 = \frac{E^2}{L^2} - (1 - 2M\tilde{G}u)\left(\frac{1}{L^2} + u^2\right). \tag{7.36}$$

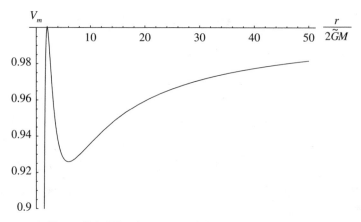

Figure 7.1. Effective potential for a massive particle.

Proceeding similarly, the orbit equation for a zero-mass particle is

$$\left(\frac{du}{d\phi}\right)^2 = \frac{E^2}{L^2} - \left(1 - 2M\tilde{G}u\right)u^2. \tag{7.37}$$

We can get insight into the nature of the possible orbits of both massive and zero-mass particles by considering Eqs. (7.32) and (7.33) as "energy" conservation equations for one-dimensional motion with effective potentials V_m and V_0 for massive and zero-mass particles, respectively:

$$V_m = \left(1 + \frac{L^2}{r^2}\right)\left(1 - \frac{2\tilde{G}M}{r}\right) \tag{7.38}$$

$$V_0 = \frac{L^2}{r^2}\left(1 - \frac{2\tilde{G}M}{r}\right). \tag{7.39}$$

An effective potential V_m is shown in Figure 7.1 with $L = 4\tilde{G}M$. Note that turning points of orbits occur at values of r such that $E^2 = V_m(r)$. We see that there are two orbits of fixed radius occurring at the extrema of V_m, V_m^a, and V_m^b, the inner corresponding to an unstable orbit of radius $a = 4\tilde{G}M$ and the outer a stable orbit of radius $b = 12\tilde{G}M$. For $V_m^b < E^2 < 1$ there exists an orbit with two turning points between $r = a$ and $r = b$ and an orbit with one turning point at a point $r < a$. For this case $(L = 4\tilde{G}M)$, all unbound orbits, which have $E^2 > 1$, have no turning points. However, $V_m^a > 1$ if $L > 4\tilde{G}M$ and unbound orbits with one turning point exist.

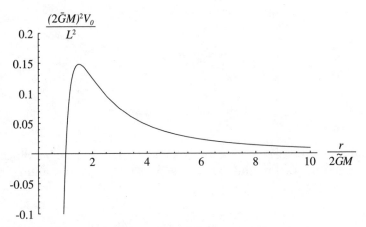

Figure 7.2. Effective potential for a zero–mass particle.

The effective potential V_0 is shown in Figure 7.2. One unstable orbit of fixed radius occurs at the single extremum of V_0, which occurs at $r = 3\tilde{G}M$. For $V_0(3\tilde{G}M) > E^2 > 0$, there is an unbound orbit with one turning point. If $E^2 > V_0(3\tilde{G}M)$, the orbit has no turning points.

7.6.1   Gravitational Deflection of Light

In Chapter 5, we saw that Einstein, in his 1911 paper, predicted a deflection of .875 arcsec for light passing near the sun. The prediction arising from his general theory is not the same, as one can realize by noting that both g_{rr} and g_{00} are changed from their flat value, not just g_{00}, as is effectively assumed in the 1911 paper. The deflection is calculated by calculating the change in $\phi, \delta\phi$, as the photon comes in from $r = \infty$, passes near the massive body at some closest point, and then proceeds out to $r = \infty$. The deflection $\Delta\phi$ is $\delta\phi - \pi$. (See Fig. 7.3.) Equivalently, we could calculate the change in ϕ as the photon proceeds in from $r = \infty$ to the point of closest approach, subtract $\pi/2$, and double the result.

Note that the point of closest approach $r_0 = 1/u_0$ is determined by setting the right-hand side of the orbit equation, Eq. (7.37), equal to zero. This determines E^2/L^2 in terms of u_0 as

$$\frac{E^2}{L^2} = \left(1 - 2M\tilde{G}u_0\right)u_0^2, \qquad (7.40)$$

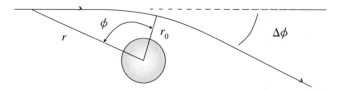

Figure 7.3. Light deflection.

and, then, the orbit equation can be written in terms of u_0. Since we want to integrate to find the change in ϕ, we write the resulting orbit equation as

$$\frac{d\phi}{du} = \pm \left(\left(1 - 2M\tilde{G}u_0 \right) u_0^2 - \left(1 - 2M\tilde{G}u \right) u^2 \right)^{-1/2}. \qquad (7.41)$$

The plus obtains for an incoming photon with $L > 0$. Thus, we find

$$\frac{\Delta\phi}{2} = \int_{u_0}^0 \frac{du}{\left(\left(1 - 2M\tilde{G}u_0 \right) u_0^2 - \left(1 - 2M\tilde{G}u \right) u^2 \right)^{1/2}} - \frac{\pi}{2}. \qquad (7.42)$$

The integral to be evaluated is an elliptic integral.

To obtain an analytic expression for the deflection angle in terms of the point of closest approach u_0, we do a weak field expansion of Eq. (7.42) using the small parameter $\tilde{G}Mu$. There are some subtleties in such an expansion. The expression for solving E^2/L^2 in terms of u_0 is linear in $\tilde{G}Mu$. Thus, in the integrand of Eq. (7.42) the small expansion parameter occurs in two forms, $\tilde{G}Mu_0$ and $\tilde{G}Mu$. Expansion of Eq. (7.42) to first order in these two parameters leads to

$$\frac{\Delta\phi}{2} \approx \int_{u_0}^0 \frac{du}{(u_0^2 - u^2)^{1/2}} + M\tilde{G} \int_{u_0}^0 \frac{u_0^3 - u^3}{(u_0^2 - u^2)^{3/2}}\, du - \frac{\pi}{2}. \qquad (7.43)$$

The first integral gives $\pi/2$. The second integral gives $2\tilde{G}Mu_0$. Thus, we have

$$\frac{\Delta\phi}{2} \approx 2\tilde{G}Mu_0 = \frac{2\tilde{G}M}{r_0}, \qquad (7.44)$$

which yields a predicted deflection of $\Delta\phi = 1.75$ arcsec, twice that predicted by Einstein in his 1911 paper.

The deflection was measured by Dyson and Eddington (Dyson et al. 1920) during a solar eclipse in 1919, yielding a result $\Delta\phi = 1.90 \pm 0.16$ arcsec, confirming the prediction within about 30% accuracy.

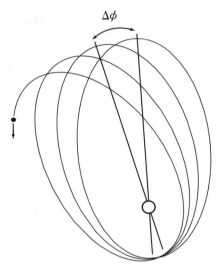

Figure 7.4. Perihelion precession.

7.6.2   Precession of the Perihelia of Orbits

In Newtonian dynamics all bound orbits for an attractive $1/r$ potential and an isotropic simple harmonic oscillator close; that is, the period of motion in the r variable is the same as that in the angle ϕ variable. For other spherically symmetric potentials this is generally not the case.

The "time" it takes for the orbiting particle to go from perihelion r_- to aphelion r_+ and back to perihelion is not the same as the time it takes the particle's angle position ϕ to change by 2π. The orbits precess. (See Fig. 7.4.)

Now consider the orbit equations of a particle about a spherical mass as given by Eqs. (7.27) and (7.32). The effective potential V_m, Eq. (7.38), is not the effective potential of a $1/r$ potential (nor, for that matter, is it one of an isotropic simple harmonic oscillator). Thus, one would expect that the orbits would precess. We will calculate the precession per orbit expressed in terms of r_- and r_+.

First, we show that the Newtonian orbit does not precess. Note that since the equivalent one-dimensional potential for the Kepler problem does not contain a $1/r^3$ term, one would expect to obtain the Newtonian orbit equation if one neglects the $1/r^3$ term in Eq. (7.32). Indeed, if this term is dropped and the equation is multiplied by $m/2$, the following equation results:

$$\frac{m}{2}\left(\frac{dr}{d\tau}\right)^2 = \left(E^2 - 1\right)m - \left(\frac{L^2 m^2}{2mr^2} - \frac{\tilde{G}mM}{r}\right).$$

For low velocity and a weak field $d\tau \approx dt$. Thus we see that the equivalent one-dimensional motion is that of the Kepler problem. Neglecting the corresponding u^3 in Eq. (7.37), we obtain for the orbit equation

$$\left(\frac{du}{d\phi}\right)^2 = \frac{E^2}{L^2} - \left(1 - 2M\tilde{G}u\right)\frac{1}{L^2} - u^2. \tag{7.45}$$

The aphelion, $u_+ = 1/r_+$, and perihelion, $u_- = 1/r_-$, are determined by the condition $du/d\phi = 0$. Solving the resulting two simultaneous equations,

$$\frac{E^2}{L^2} - \left(1 - 2M\tilde{G}u_\pm\right)\frac{1}{L^2} - u_\pm^2 = 0,$$

for E^2/L^2 and $1/L^2$, one obtains

$$\frac{1}{L^2} = \frac{u_+ + u_- - 2\tilde{G}Mu_+ u_-}{2\tilde{G}M}, \tag{7.46}$$

and

$$\frac{E^2}{L^2} = \frac{u_+ + u_-}{2\tilde{G}M}. \tag{7.47}$$

With these expressions, Eq. (7.45) becomes

$$\left(\frac{du}{d\phi}\right)^2 = \left(u_- - u\right)\left(u - u_+\right), \tag{7.48}$$

and thus

$$\frac{d\phi}{du} = \frac{1}{\left(u_- - u\right)^{1/2}\left(u - u_+\right)^{1/2}}. \tag{7.49}$$

As the particle proceeds from aphelion to perihelion, the change in ϕ is thus

$$\phi(u_-) - \phi(u_+) = \int_{u_+}^{u_-} \frac{du}{\left(u_- - u\right)^{1/2}\left(u - u_+\right)^{1/2}} = \pi. \tag{7.50}$$

Since the change is the same as the particle proceeds from perihelion to aphelion, the total change in ϕ is 2π; the orbit closes.

It is clear how to calculate the precession using the orbit equation Eq. (7.36). Express the right-hand side in terms of u_+ and u_-, obtained by setting the right-hand side equal to zero; integrate the equation for $d\phi/du$ from u_+ to u_- and double the result. The result is the precession mod 2π. The integral to be evaluated is the elliptic integral

$$\phi(u_-) - \phi(u_+) = \int_{u_+}^{u_-} \frac{du}{\left[\frac{E^2}{L^2} - (1 - 2M\tilde{G}u)\left(\frac{1}{L^2} + u^2\right)\right]^{1/2}}. \quad (7.51)$$

As noted, the expressions for E^2/L^2 and $1/L^2$ in terms of u_+ and u_- are obtained by solving the simultaneous equations that result from setting the right-hand side of Eq. (7.37) equal to zero:

$$\frac{E^2}{L^2} - (1 - 2M\tilde{G}u_\pm)\left(\frac{1}{L^2} + u_\pm^2\right) = 0.$$

(If one views these equations as equations for $u_\pm$ in terms of E^2/L^2 and $1/L^2$, they are cubic and generally yield three values for $u > 0$. We expect this from the above discussion of the nature of orbits. However, any two such solutions determine E^2/L^2 and $1/L^2$.)
    We obtain

$$\frac{E^2}{L^2} = \frac{u_+ + u_-}{2M\tilde{G}} - (u_+^2 + u_-^2 + 2u_+ u_-) + 2M\tilde{G}(u_+^2 u_- + u_-^2 u_+), \quad (7.52)$$

and

$$\frac{1}{L^2} = \frac{u_+ + u_-}{2M\tilde{G}} - (u_+^2 + u_-^2 + u_+ u_-). \quad (7.53)$$

The substitution of these expressions into Eq. (7.51) gives
$\phi(u_-) - \phi(u_+) =$

$$\int_{u_+}^{u_-} \frac{du}{\left((u - u_+)(u_- - u) - 2M\tilde{G}(u - u_-)(u_+ - u)(u_+ + u_- + u)\right)^{1/2}}. \quad (7.54)$$

As expected, setting the term proportional to $2M\tilde{G}$ to zero gives the Newtonian result Eq. (7.49). Treating this term to first order, however, gives

$$\phi(u_-) - \phi(u_+) \approx$$

$$\int_{u_+}^{u_-} \frac{du}{(u_- - u)^{1/2}(u - u_+)^{1/2}} + M\tilde{G} \int_{u_+}^{u_-} \frac{(u_+ + u_- + u)\,du}{(u_- - u)^{1/2}(u - u_+)^{1/2}}. \quad (7.55)$$

The first integral gives π, the Newtonian result, whereas the second gives $3M\tilde{G}(u_+ + u_-)\,\pi/2$. The change in ϕ as the particle moves from aphelion to perihelion and back to aphelion is $2\pi + 3\pi\tilde{G}M(u_+ + u_-)$; the orbit precesses by an angle $\Delta\phi$ given by

$$\Delta\phi = 3\pi\tilde{G}M(u_+ + u_-) = 3\pi\tilde{G}M\left(\frac{1}{r_+} + \frac{1}{r_1}\right). \tag{7.56}$$

The perihelion advances by this amount for each orbit as depicted in Figure 7.4.

For Mercury, the prediction is a perihelion advance of 43 arcsec per century, a prediction experimentally confirmed by observation. The actual observed advance must be corrected by effects of planetary perturbations to effect the agreement with the predicted advance.

7.7 Orbiting Gyroscopes in General Relativity

We argued in Chapter 3, Eq. (3.20), that a gyroscope, in special relativity, obeys the dynamical equation

$$\frac{dL^\alpha(\tau)}{d\tau} = -L^\sigma n_{\sigma\beta}\frac{dU^\beta}{d\tau}U^\alpha. \tag{7.57}$$

By the *principle of general covariance*, in curved space-time this equation becomes

$$\frac{DL^\alpha(\tau)}{d\tau} = -L^\sigma g_{\sigma\beta}\frac{DU^\beta}{d\tau}U^\alpha, \tag{7.58}$$

obtained merely by replacing the ordinary derivatives by covariant derivatives. For an orbiting gyroscope,

$$\frac{DU^\beta}{d\tau} = 0, \tag{7.59}$$

which is the geodesic equation, and Eq. (7.56) becomes

$$\frac{DL^\alpha(\tau)}{d\tau} = 0, \tag{7.60}$$

or, written out,

$$\frac{dL^\alpha}{d\tau} + \Gamma^\alpha_{\sigma\beta}L^\sigma U^\beta = 0. \tag{7.61}$$

Note that this equation merely states that the four-vector L^α is parallel-transported (in space-time) along the orbit. We might then expect that the value of the four-vector, when the gyroscope is at a point on a closed orbit, will not repeat when the gyroscope returns to this point. The gyroscope is expected to precess, meaning the "direction" of the spin of the gyroscope is different when it returns to its starting point from what it was at the start. One can observe the direction the gyroscope points with respect to the light received from fixed stars at the beginning of an orbit and at the completion. A comparison would give the angle of precession.

If we write $L^\alpha(2\pi)$ for the value of L^α when the gyroscope completes the orbit and L_0^α for the value at the beginning of the orbit, the precession angle, δ_p, after the completion of one orbit, satisfies the relation

$$\cos\delta_p = \frac{\mathbf{L}_0 \cdot \mathbf{L}(2\pi)}{\mathbf{L}_0 \cdot \mathbf{L}_0},$$

where $\mathbf{L}$ is the three-vector angular momentum in the rest frame of the gyroscope. Recall that in the rest frame of the gyroscope the time-component of the four-vector L^α is zero. Thus, we can rewrite this relation for δ_p in the covariant form,

$$\cos\delta_p = \frac{L_0^\alpha L_\alpha(2\pi)}{L_0^\alpha L_{0\alpha}}. \tag{7.62}$$

We calculate this precession for a gyroscope in a circular orbit with $\theta = \pi/2$ in the Schwarzschild metric, eventually restricting to weak fields and small velocity. First, we write, in terms of ϕ rather than τ derivatives,

$$\frac{dL^\alpha}{d\phi} + \Gamma^\alpha_{\sigma\beta} L^\sigma \frac{dx^\beta}{d\phi} = 0. \tag{7.63}$$

Furthermore, from Eqs. (7.25) and (7.27),

$$\frac{dx^\sigma}{d\phi} = \frac{dx^\sigma}{d\tau}\frac{d\tau}{d\phi} = \left(\frac{E}{L}\frac{r^2}{1 - 2\tilde{G}M/r}, 0, 0, 1\right). \tag{7.64}$$

For a circular orbit, $u_+ = u_- = 1/r$, Eq. (7.52) gives

$$\frac{E}{L} = \frac{1 - 2\tilde{G}M}{(\tilde{G}Mr)^{1/2}}. \tag{7.65}$$

With the Schwarzschild metric's Christoffel symbols (Sec. 7.3), Eqs. (7.64) and (7.65), Eq. (7.63) becomes

$$\frac{dL^\alpha}{d\phi} = -\begin{pmatrix} \left(\frac{\tilde{G}M}{r}\right)^{1/2} \frac{L^r}{1-2\tilde{G}M/r}, \left(\frac{\tilde{G}M}{r}\right)^{1/2}(1-2\tilde{G}M/r)L^0 \\ -(1-2\tilde{G}M/r)rL^\phi, 0, \frac{L^r}{r} \end{pmatrix}. \tag{7.66}$$

From this, we find that

$$\frac{d^2 L^r}{d\phi^2} = -\left(1 - \frac{3\tilde{G}M}{r}\right)L^r. \tag{7.67}$$

Let us consider the case for which $L^\alpha(\phi = 0) = (0, L_0^r, 0, 0)$. Clearly, $L^\alpha(0)U_\alpha(0) = 0$, as it must. Also, from Eq. (7.66), $dL^r/d\phi = 0$ at $\phi = 0$. With these conditions the solution to Eq. (7.67) is

$$L^r = L_0^r \cos\left[\left(1 - \frac{3\tilde{G}m}{r}\right)^{1/2}\phi\right], \tag{7.68}$$

and thus

$$L^r(2\pi) = L_0^r \cos\left[\left(1 - \frac{3\tilde{G}m}{r}\right)^{1/2}2\pi\right]. \tag{7.69}$$

One could use this solution in Eq. (7.65) and solve for the ϕ dependence of L^0 and L^ϕ. However, to solve for the amount the gyroscope precesses in one orbit, we can use Eq. (7.69) in Eq. (7.62) with the result

$$\delta_p = \left(1 - \frac{3\tilde{G}M}{r}\right)^{1/2}2\pi - 2\pi \approx \frac{3\tilde{G}M\pi}{r}. \tag{7.70}$$

This is referred to as the *de Sitter precession*. It is not the analog of the Thomas precession—rather, it is the precession of freely falling frames, not frames that are accelerated, as is the case of the Thomas precession. Note that the weak-field result, $3\tilde{G}M\pi/r$, is the precession of the perihelion of an orbit in the limit of a circular orbit.

For a gyroscope orbiting around a spinning body, such as the earth, for which the metric is not quite spherically symmetric, there is an additional contribution to the precession due to "frame dragging." This additional precession is called the *Lense-Thirring precession*. In order to discuss this effect, we would have to derive the modification of the metric due to rotation, something we will not do.

Both the de Sitter and the Lense-Thirring precession are to be measured by Stanford's Gravity B Probe, scheduled for launch in 2003.

7.8 Stellar Interiors

We have obtained the metric outside of a static spherical distribution of matter and will now study the equations for a static spherical metric in the presence of matter. Such metrics are the metrics of the interior of nonrotating stars and must smoothly go over to the Schwarzschild solution at the surface of the material.

The metric is determined, through the Einstein field equations, by the properties of the matter as expressed by the energy-momentum tensor, which we assume to be that of a perfect fluid, and the equation of state of the fluid, which relates the pressure of the fluid to its energy density.

Since the metric is assumed to be static and spherically symmetric, the invariant interval can again be put in the form of Eq. (7.2):

$$d\tau^2 = B(r)(dx^0)^2 - A(r)dr^2 - r^2 d\theta^2 - r^2 \sin^2\theta d\phi^2. \tag{7.71}$$

This, of course, means that we again have

$$R_{00} = -\frac{B'(r)}{rA(r)} + \frac{1}{4}\left(\frac{B'(r)}{A(r)}\right)\left(\frac{A'(r)}{A(r)} + \frac{B'(r)}{B(r)}\right) - \frac{B''(r)}{2A(r)} \tag{7.72}$$

$$R_{rr} = -\frac{A'(r)}{rA(r)} - \frac{1}{4}\left(\frac{B'(r)}{B(r)}\right)\left(\frac{A'(r)}{A(r)} + \frac{B'(r)}{B(r)}\right) + \frac{B''(r)}{2B(r)} \tag{7.73}$$

$$R_{\theta\theta} = -1 + \frac{r}{2A(r)}\left(-\frac{A'(r)}{A(r)} + \frac{B'(r)}{B(r)}\right) + \frac{1}{A(r)} \tag{7.74}$$

$$R_{\phi\phi} = \sin^2(\phi) R_{\theta\theta} \tag{7.75}$$

$$R_{\mu\nu} = 0; \quad \mu \neq \nu. \tag{7.76}$$

Since the metric is static and spherically symmetric, the energy-momentum tensor must also be. Thus,

$$T^{\alpha\beta} = -p(r)g^{\alpha\beta} + (p(r) + \rho(r))U^\alpha(r)U^\beta(r), \tag{7.77}$$

with $U^\alpha = U^0(r)\delta^{0\alpha}$. The four-velocity has a zero component only. (One might think it could have a nonvanishing r component, but if it did T^{0r} would be nonvanishing, whereas $R^{0r} = 0$ implies it vanishes.) Further, since $U^\alpha U^\beta g_{\alpha\beta} = U^0(r)U^0(r)g_{00}(r) = 1$ and $U^0(r)U^0(r) = B^{-1}(r)$, the energy-momentum tensor, in covariant form, becomes

$$T_{\alpha\beta} = -p(r)g_{\alpha\beta} + (p(r) + \rho(r))\delta_{0\alpha}\delta_{0\beta}B(r). \qquad (7.78)$$

For a perfect fluid $T = T_\beta^\beta = \rho(r) - 3p(r)$. Thus, we have

$$T_{\alpha\beta} - g_{\alpha\beta}T/2 = \frac{p-\rho}{2}g_{\alpha\beta} + (p(r) + \rho(r))\delta_{0\alpha}\delta_{0\beta}B(r). \qquad (7.79)$$

With this and Eqs. (7.72) to (7.76), Einstein's field equation, Eq. (6.70), becomes

$$-\frac{B'(r)}{rA(r)} + \frac{1}{4}\left(\frac{B'(r)}{A(r)}\right)\left(\frac{A'(r)}{A(r)} + \frac{B'(r)}{B(r)}\right) - \frac{B''(r)}{2A(r)} = -8\pi\tilde{G}\frac{3p+\rho}{2}B(r) \qquad (7.80)$$

$$-\frac{A'(r)}{rA(r)} - \frac{1}{4}\left(\frac{B'(r)}{B(r)}\right)\left(\frac{A'(r)}{A(r)} + \frac{B'(r)}{B(r)}\right) + \frac{B''(r)}{2B(r)} = 8\pi\tilde{G}\frac{p-\rho}{2}A(r) \qquad (7.81)$$

$$-1 + \frac{r}{2A(r)}\left(-\frac{A'(r)}{A(r)} + \frac{B'(r)}{B(r)}\right) + \frac{1}{A(r)} = 8\pi\tilde{G}\frac{p-\rho}{2}r^2. \qquad (7.82)$$

The local conservation of energy-momentum equation, $T^{\alpha\beta}{}_{;\beta} = 0$, gives

$$T^{r\beta}{}_{;\beta} = \frac{dp(r)}{dr}A^{-1}(r) + (p+\rho)A^{-1}(r)\frac{B'(r)}{2B(r)} = 0. \qquad (7.83)$$

(The other components of the equation are trivially satisfied.) The computation of this covariant derivative requires knowledge of the Christoffel symbols for the metric of Eq. (7.71). (See Exercise 7.1.) Eq. (7.83) is not independent of Eqs. (7.80)–(7.82) but is satisfied as a result of the Bianchi identity. However, it has a particularly useful form.

To solve for the metric, the procedure will be to use Eqs. (7.80)–(7.82) to obtain a solution for $A(r)$ and $B(r)$ in terms of a $p(r)$ and $\rho(r)$, and then Eq. (7.83) will be a first-order differential equation relating $p(r)$ to $\rho(r)$, to be solved consistent with the assumed equation of state. To this end we see that a rather simple combination of Eqs. (7.80)—(7.82) yields an equation for $A(r)$,

$$\frac{R_{00}}{2B(r)} + \frac{R_{rr}}{2A(r)} + \frac{R_{\theta\theta}}{r^2} = -\frac{1}{r^2} - \frac{A'(r)}{rA^2(r)} + \frac{1}{r^2A(r)} = -8\pi\tilde{G}\rho, \qquad (7.84)$$

which can be rewritten as

$$\frac{d(rA^{-1}(r))}{dr} = -8\pi\tilde{G}\rho r^2 + 1. \qquad (7.85)$$

Equation (7.85) has a solution

$$A(r) = \left(1 - \frac{2\tilde{G}m(r)}{r} + \frac{c}{r}\right)^{-1}. \tag{7.86}$$

Here

$$m(r) = \int_0^r 4\pi r'^2 \rho(r')\,dr', \tag{7.87}$$

and c is an integration constant. Since the space-space part of the metric must become a three-dimensional Euclidean space for small r, c must be zero. With $A(r)$ known, Eq. (7.82) can be solved for B'/B with the result

$$\frac{B'(r)}{B(r)} = \frac{2\tilde{G}\left[4\pi pr^3 + m(r)\right]}{r\left[r - 2\tilde{G}m(r)\right]}. \tag{7.88}$$

This use in Eq. (7.83) results in the following equation, known as the *Oppenheimer-Volkoff (O-V) equation:*

$$\frac{dp(r)}{dr} = -(p + \rho)\frac{\tilde{G}\left[4\pi pr^3 + m(r)\right]}{r\left[r - 2\tilde{G}m(r)\right]}. \tag{7.89}$$

Integration of the O-V equation, consistent with the equation of state relating ρ to p, with some assumed value of $p(0)$, will give $p(r)$ and $\rho(r)$. Note that $p(r)$ is monotonically decreasing with increasing r if $\rho \geq 0$ and $2\tilde{G}m(r) < r$. Also note that the expression for $A(r)$ becomes singular if $2\tilde{G}m(r) = r$. It is, thus, reasonable to assume that $2\tilde{G}m(r) < r$ in the interior of a *static* star. (See the next section.) With this condition, the pressure monotonically decreases from its central value until it reaches zero at some value of r, denoted by r_s, the "radius" of the star.[2] The value of $m(r_s)$ equals the "mass" parameter M of the Schwarzschild metric for the star's exterior. We see that the "radius" and the "mass" of the star are determined by the central pressure p_c and, of course, the equation of state.

One should not make too much of the relation

$$m(r_s) = \int_0^{rs} 4\pi r^2 \rho(r)\,dr = M,$$

a relation that is identical to that in Newtonian theory. Remember that $\rho(r)$ is the energy density in an inertial coordinate system at rest with respect to the fluid, and the coordinates $\{x^0, r, \theta, \phi\}$ are not such

---

[2] With the reasonable assumption that the equation of state is such that $dp/d\rho > 0$, the energy density also monotonically decreases from its central value.

coordinates. More to the point, $4\pi r^2\, dr$ is not a "proper" volume element. After all, the proper distance, at fixed coordinate time, between r and $r + dr$ is $(1 - \tilde{G}m(r)/r)^{-1/2}\, dr$, not dr. Thus, a proper volume, at fixed coordinate time, is

$$\left(1 - \frac{\tilde{G}m(r)}{r}\right)^{-1/2} r^2\, dr d\Omega,$$

not $r^2\, dr d\Omega$. The integral cannot be identified with the total energy of the star, but only with the "mass" parameter of the exterior Schwarzschild metric.

In the Newtonian limit with $p \ll \rho$ and $\tilde{G}m(r) \ll r$, the O-V equation becomes

$$\frac{dp}{dr} \approx -\frac{\tilde{G}m(r)}{r^2}, \tag{7.90}$$

the Newtonian hydrostatic equilibrium equation. An interesting difference in the stellar equilibrium equations of general relativity and Newtonian gravity is seen by comparing Eq. (7.90) with O-V equation for a given density distribution $\rho(r)$. It is easy to see that for this case $|dp/dr|$ is greater for the O-V equation than it is for the Newtonian Eq. (7.90). This in turn implies that the pressure at each r is greater in the general relativistic star than in the Newtonian star. The relativistic star requires a higher pressure than a Newtonian star to keep the material from collapsing. A significant implication of this is well illustrated by considering a star made of an incompressible fluid, that is, one for which $\rho(r)$ is a constant ρ_0. In this case, for both type of stars

$$m(r) = \frac{4\pi}{3} r^3 \rho_0 = M \frac{r^3}{r_s^3}, \tag{7.91}$$

where M is the "mass" parameter of the exterior Schwarzschild metric and r_s is the "radius" of the star—the "radius" at which the pressure vanishes.

7.8.1 Constant Density Newtonian Star

The Newtonian equilibrium equation, Eq. (7.90), by use of Eq. (7.91) becomes

$$\frac{dp}{dr} = -\frac{\tilde{G}M}{r_s^3} r, \tag{7.92}$$

which is easily integrated to yield

$$p = \frac{\tilde{G}M}{2r_s}\left[1 - \left(\frac{r}{r_s}\right)^2\right]. \tag{7.93}$$

The condition $p(r_s) = 0$ has been imposed. The central pressure p_s is $\tilde{G}M/2r_s$ and is finite for all M and r_s.

7.8.2 Constant Density Relativistic Star

The O-V equation takes the form

$$\frac{dp(r)}{dr} = (p + \rho_0)(3p + \rho_0)\frac{-4\pi\tilde{G}r}{3 - 8\pi\tilde{G}r^2\rho_0}, \tag{7.94}$$

or equivalently,

$$\frac{dp}{(p + \rho_0)(3p + \rho_0)} = \frac{-4\pi\tilde{G}rdr}{3 - 8\pi\tilde{G}r^2\rho_0}. \tag{7.95}$$

This last equation, upon integration, yields

$$\frac{3p + \rho_0}{p + \rho_0} = \left(\frac{3 - 8\pi\tilde{G}\rho_0 r^2}{3 - 8\pi\tilde{G}\rho_0 r_s^2}\right)^{1/2}. \tag{7.96}$$

Again, the condition $p(r_s) = 0$ has been applied. Solving for $p(r)$, we have

$$p(r) = -\rho_0\frac{(3 - 8\pi\tilde{G}\rho_0 r_s^2)^{1/2} - (3 - 8\pi\tilde{G}\rho_0 r^2)^{1/2}}{3(3 - 8\pi\tilde{G}\rho_0 r_s^2)^{1/2} - (3 - 8\pi\tilde{G}\rho_0 r^2)^{1/2}}. \tag{7.97}$$

This implies an upper limit on the density ρ_0 for a star of a fixed-radius star. This upper-limit density is that for which the central pressure becomes infinite, which occurs when the denominator of the right-hand side of Eq. (7.97) vanishes. Thus,

$$8\pi\tilde{G}r_s^2\rho_0 < \frac{8}{3}. \tag{7.98}$$

Expressed in terms of the mass M of the star, this inequality becomes

$$\frac{\tilde{G}M}{r_s} < \frac{4}{9}. \tag{7.99}$$

A static homogeneous star made of an incompressible fluid cannot exceed this mass -to- radius ratio. Since incompressible fluid has the "stiffest" equation of state possible, one would expect that the inequality Eq. (7.99) would hold for any equation of state. That such is the case was proved by Buchdahl (Buchdahl 1959). It should be noted that an incompressible fluid is unphysical. The velocity of sound in a perfect fluid is given by $\partial p/\partial \rho$, which for an incompressible fluid is infinite.

We can complete the solution for the metric of this relativistic homogeneous star by solving for $B(r)$ using Eq. (7.83) written in the form

$$\frac{B'(r)}{B(r)} = -\frac{2p'(r)}{p(r)+\rho_0}. \tag{7.100}$$

Upon integration, this yields

$$B(r) = \frac{\rho_0^2\left(1 - 2\tilde{G}M/r_s\right)}{(\rho_0 + p(r))^2}. \tag{7.101}$$

The boundary condition $B(r_s) = 1 - 2\tilde{G}M/r_s$ has been imposed. This assures that the internal metric matches the external metric at the star's surface. Finally, with $p(r)$ given by Eq. (7.97), we obtain

$$B(r) = \frac{1}{4}\left[3\left(1 - \frac{2\tilde{G}M}{r_s}\right)^{1/2} - \left(1 - \frac{2\tilde{G}Mr}{r_s^2}\right)^{1/2}\right]^2. \tag{7.102}$$

It is noteworthy that the bound on $\tilde{G}/r_s$ given by Eq. (7.99) restricts the size of the gravitational redshift of light emitted from the surface of a static spherical star. From Eq. (7.19), for the *fractional* increase in the wavelength, denoted by z, we have

$$z = \frac{\lambda_2 - \lambda_1}{\lambda_1} = \frac{\lambda_2}{\lambda_1} - 1 = \frac{g_{00}(r_2)^{1/2}}{g_{00}(r_1)^{1/2}} - 1 = B(r_1)^{-1/2} - 1. \tag{7.103}$$

Here r_1 is the radial position at which the light is emitted and, it is assumed, the light is received at $r = \infty$. Thus, a star satisfying the bound of Eq. (7.99) has a redshift parameter z less than two for light originating at the surface. Of course, light coming from the interior of the star, imagined opaque, does not respect this bound. In fact, for a star saturating the bound, $B(0) = \infty$ and thus z is unbounded.

7.9 Black Holes

It seems something peculiar occurs to the Schwarzschild metric, Eq. (7.18), at $r = r_H \equiv 2\tilde{G}M$. The coefficient of $(dx^0)^2$ becomes zero—the invariant interval becomes independent of dx^0. In addition, the coefficient of dr^2 becomes infinite so that a small dr causes the $d\tau$ to blow up. The metric has a very singular behavior here. Is this singular

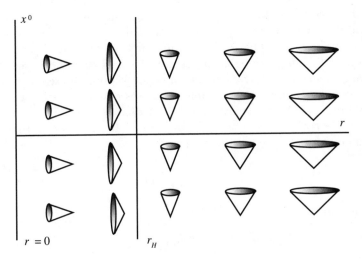

Figure 7.5. Schwarzschild light cones.

behavior intrinsic to the space-time geometry or is it merely reflecting properties of the space-time coordinates being used?

Consider the metric of a sphere expressed in terms of polar and azimuthal angles,

$$dl^2 = d^2\theta + \sin^2\theta d^2\phi.$$

At the pole, the coefficient of $d^2\phi$ vanishes; however, we know there is nothing peculiar about the geometry of the sphere at a pole—the geometry of a sphere is the same at all points. This singular behavior is a property of the coordinates, not an intrinsic property of the sphere.

Though the Schwarzschild metric is not well behaved at $r = r_H$, and some components of the Riemannian curvature tensor are singular at $r = r_H$ (see Exercise 2), the metric does satisfy the vacuum field equations $R^{\alpha\beta} = 0$ for both $r > r_H$ and $r < r_H$. However, since $g_{00} < 0$ and $g_{rr} > 0$ for $r < r_H$ in this region, x^0 is a spacelike coordinate and r is a timelike variable. We obtain a good picture of this geometry by considering the light cones associated with "radially" traveling light defined by

$$d\tau^2 = \left(1 - \frac{r_H}{r}\right)(dx^0)^2 - \left(1 - \frac{r_H}{r}\right)^{-1}(dr)^2 = 0.$$

These cones are depicted in Figure 7.5. Remembering that r is a "time" variable for $r < r_H$ the metric is not "stationary" that is, it is not r independent.

If we could perform a change of coordinates such that the metric and the Riemann tensor are well behaved at $r = r_H$, we would know that the singular behavior is not intrinsic to the metric. A manifestation of the singular behavior of the metric is that the radial light cones close up as $r \to r_H$. One can imagine keeping these cones open by choosing a new time variable x'^0 that is r dependent, so that the ingoing radial rays keep slope -1. With a change of coordinates $x^0 = x'^0 + f(r'), r = r', \theta' = \theta, \phi' = \phi$ for $r > r_H$, the condition that the ingoing radial cone has slope -1 is

$$\frac{df}{dr'} = -\left(\frac{r'}{r_H} - 1\right)^{-1},$$

which gives

$$f(r') = -r_H \ln\left(\frac{r'}{r_H} - 1\right), \quad r' > r_H.$$

The transformed Schwarzschild invariant interval in these coordinates, called the *Eddington-Finkelstein coordinates*, becomes

$$d\tau^2 = \left(1 - \frac{r_H}{r}\right)(dx'^0)^2 - \left(1 + \frac{r_H}{r}\right)dr^2 - 2\frac{r_H}{r}dx^0 dr - r^2 d\theta^2 - r^2 \sin^2\theta d\phi^2.$$

$$(7.104)$$

Though the coordinate change was meaningful only for $r > r_H$, the resulting metric is well behaved for $r > 0$, the vacuum field equations are satisfied for $r > 0$, and the components of the Riemann tensor are well behaved for $r > 0$ (Exercise 3). Using Eq. (7.104), we see that radially "outgoing" light has a slope given by

$$\frac{dx'^0}{dr} = \frac{r + r_H}{r - r_H}. \qquad (7.105)$$

For $r > r_H$ this slope is in fact positive, but for $r < r_H$ it is negative. For $r < r_H$, both radially proceeding light rays move inward. The resulting light cones are depicted in Figure 7.6.

Light cannot proceed outward originating at $r < r_H$. Since the world lines of particles of nonzero mass move within the future light cone, such particles cannot pass outward past r_H, called the *Schwarzschild radius*. The "sphere" defined by $r = r_H$ is referred to as the black hole's *event horizon*. If the mass of an object such as a star becomes contained in this sphere, a black hole is formed. The ultimate fate of a star depends on the dynamics of the material forming the star. Though the exterior solution is a Schwarzschild metric, the interior solution,

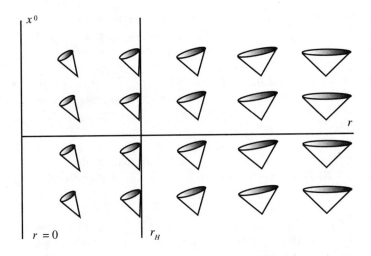

Figure 7.6. Future light cones in Eddington-Finkelstein coordinates.

and thus the development of the surface that, as described in the last section, is characterized by the radius at which the density and pressure vanish, depends on the details of the dynamics.

Nevertheless, we can question the ultimate fate of a spherical star if its surface, moving inward, passes the radius of the Buchdal bound, Eq. (7.99), $r_s = 9\tilde{G}M/4$. Since the bound would be exceeded, the material cannot return to a static distribution. Rather, it must continue to flow inward. But to what end? The metric exterior to the matter remains Schwarzschild as follows from Birkhoff's theorem. Thus, eventually the surface of the star must pass the Schwarzschild radius. A black hole, of necessity, forms.

A realistic study of the dynamics of a star is required to see under what conditions it would evolve so that a black hole is indeed formed. Such studies require knowledge of the equation of state of material consisting of nuclei and other elementary particles under conditions for which quantum effects are important. This we will not consider.

7.10   Exercises

1. Starting with the metric of the form Eq. (7.2), compute (a) the Christoffel symbols and (b) the Ricci tensor to obtain Eqs. (7.3) to (7.7).

2. (a) Using Schwarzschild coordinates, compute the Schwarzschild metric's Christoffel symbols. (b) Compute the metric's Riemann tensor. Are any of the elements singular? If so, where?

3. (a) Using Eddington-Finkelstein coordinates, compute the Schwarzschild metric's Christoffel symbols. (b) Compute the metric's Riemann tensor. Are any of its elements singular? If so, where?

4. A photon travels in the unstable circular orbit at the radius $r = 3\tilde{G}M$. What is the elapsed Schwarzschild coordinate time for one orbit?

5. A clock is in a circular orbit at $r = 12\tilde{G}M$. (a) How much time elapses on the clock during one orbit? (b) A second clock is held stationary at a point on the orbit of the first clock. What is the elapsed time the second clock reads between successive passes the of the first clock? (c) Is (a) or (b) greater? Is this what you expect? (Consider the twin "paradox".)

6. A particle is held at a fixed "radius" in Schwarzschild space. (a) What is the magnitude of the three-acceleration of the particle in an inertial frame instantaneously moving with the particle? (*Hint:* What is the covariant definition of the acceleration (i.e., what tensor reduces to the four-acceleration in an inertial frame)? (b) In a weak-field region (i.e., where $\tilde{G}M/r \ll 1$), what is the acceleration? Is this what you would expect? (c) At the Schwarzschild radius, what is the acceleration?

7. A photon propagates radially outward in a Schwarzschild metric of mass M. (a) What is the elapsed coordinate time x^0 as it moves from r_1 to r_2 ? (b) In terms of $p_0 = E$, what is the energy at r_1 as measured by an observer fixed at r_1? (*Hint:* The four-velocity of this observer $U^\alpha = dx^\alpha/d\tau$ has "space" components zero. Also, in his inertial frame $dx^i/d\tau = 0$. But $U^\alpha U_\alpha = 1$. So what is U^0? Then consider the invariant $p^\alpha U_\alpha$.) (c) Show that (b) implies the gravitational redshift given by Eq. (7.19).

8. A particle of mass m is "freely falling" outward in a Schwarzschild metric of mass M. (a) What is the value of the constant of motion E, in $p_0 = mE$, such that the particle just reaches $r = \infty$? (b) In terms of this value of E, what is the energy of the particle at $r = r_1$,

as measured by on observer fixed at r_i? (See the hint in Exercise 3.) (c) In the weak-field approximation, $\tilde{G}M/r \ll 1$, what is the approximate energy of the particle at r_1 as measured by an observer fixed at r_i? Interpret this expression in terms of the escape velocity one obtains in Newtonian theory.

9. Suppose the Einstein field equations have a nonvanishing cosmological constant λ. (a) For a static spherically symmetric metric in a vacuum, what are the equations for the metric components $A(r)$ and $B(r)$. That is, what replaces the Eqs. (7.9) to (7.12)? (b) Solve the equations of (a) and show there doesn't exist an asymptotically flat solution.

10. Derive the Eq. (7.66) for the precession of a gyroscope in circular orbit moving in a Schwarzschild metric. How does the de Sitter precession for one orbit around the earth compare to the Thomas precession one would calculate treating the gyroscope as an accelerating body moving in a circle?

Chapter 8

Metrics with Symmetry

8.1 Introduction

The symmetries of a metric are important in the study of dynamically conserved quantities and, as we will see in Chapter 9, play a defining role in the study of cosmology.[1] And, of course, the assumption of a symmetry of the metric simplifies the Einstein equations, as we saw in the discussion of the Schwarzschild metric and of the equations of motion for freely falling particles. We will formulate the meaning of a symmetry of a metric without reference to a particular coordinate system which reflects the symmetry—that is, we will characterize the the symmetry in covariant language.

This characterization can be illustrated by the consideration of the metric of a two-dimensional sphere endowed with the metric induced by imbedding in a three-dimensional Euclidean space. In the following we will use this case to illustrate symmetry concepts that have a more general applicability.

8.2 Metric Automorphisms

The two-dimensional sphere's metric, using the usual polar coordinate system, is symmetric under rotation about the polar axis—the metric is independent of the polar angle. But by considering the metric itself, how can we know it is symmetric under rotation about any direction in the embedding space? Indeed, what does it mean to say that the metric is symmetric under these rotations?

---

[1] Chapter 9 can, however, be understood without studying this chapter.

143

Under rotation each point of the sphere $x^i = (\theta, \phi)$ moves to the point $x'^j = (\theta', \phi')$. Formally, we represent this as

$$\theta' = \theta'(\theta, \phi, \boldsymbol{\delta}) \tag{8.1}$$

$$\phi' = \phi'(\theta, \phi, \boldsymbol{\delta}). \tag{8.2}$$

Here the direction and magnitude of the rotation is indicated by $\boldsymbol{\delta}$. Such a one-to-one mapping of a space to itself is called an *automorphism*. There are many automorphisms of a space, and not all concern a symmetry of the metric. The definition of an automorphism makes no reference to the metric. What characterizes automorphisms that implement symmetry transformations of the metric? The rotations of the sphere carry any differential dx^i to dx'^i in such a way that

$$g_{ij}(x') \, dx'^i \, dx'^j = g_{kl}(x) \, dx^k \, dx^l. \tag{8.3}$$

That is, any "translated" local invariant interval is unchanged. This is clearly a property of the metric and the automorphism. Eq. (8.3) can be written as

$$\left(g_{ij}(x') - g_{kl}(x) \frac{\partial x^k}{\partial x'^i} \frac{\partial x^l}{\partial x'^j} \right) dx'^i \, dx'^j = 0. \tag{8.4}$$

Since this must be true for any interval dx'^j, we have

$$g_{ij}(x') - g_{kl}(x) \frac{\partial x^k}{\partial x'^i} \frac{\partial x^l}{\partial x'^j} = 0, \tag{8.5}$$

or, equivalently,

$$g_{kl}(x) - g_{ij}(x') \frac{\partial x'^i}{\partial x^k} \frac{\partial x'^i}{\partial x^l} = 0. \tag{8.6}$$

The metric at the transformed point is related to the metric at the original point by the coordinate transformation induced by the automorphism—the automorphism is a symmetry transformation of the metric. Such an automorphism is called a *metric automorphism* or an *isometry*. We have our answer to the question of the meaning of a symmetry of a metric without reference to a particular coordinate system—the existence of a metric automorphism.

If one views the automorphism as a coordinate change at the the point x, Eq. (8.6) can be written

$$g'_{ij}(x) = g_{ij}(x), \quad \text{all} \quad x. \tag{8.7}$$

That is, the transformed metric is the same function of its arguments as the original metric is of *its* arguments. The metric is said to be *form invariant.*

8.3   Killing Vectors

Given a metric $g_{ij}(x)$, a solution, $x'^i(x)$, of the partial differential equations Eq. (8.6) exhibits an isometry of the metric. Except for special cases, these equations are complicated. (If the metric is independent of a coordinate, a simple solution of these equations can be obtained.) For this reason, it is well to consider infinitesimal isometries for which the motion of the points is small. Also, it seems reasonable that many, though perhaps not all, finite isometries could be implemented by a sequence of infinitesimal ones. Consider such an infintesimal transformation,

$$x'^i = x^i + \epsilon\, \zeta^i(x), \quad |\epsilon| \ll 1. \tag{8.8}$$

To first order in ϵ, Eq. (8.6) gives

$$\frac{\partial g_{ij}(x)}{\partial x^m}\, \zeta^m(x) + g_{il}(x)\frac{\partial \zeta^l}{\partial x^j} + g_{kj}(x)\frac{\partial \zeta^k}{\partial x^i} = 0. \tag{8.9}$$

Expressed in terms of the covariant components ζ_m, this becomes

$$
\begin{aligned}
0 &= \frac{\partial \zeta_i}{\partial x^j} + \frac{\partial \zeta_j}{\partial x^i} + \zeta^m\frac{\partial g_{ij}}{\partial x^m} - \zeta^l\frac{\partial g_{il}}{\partial x^j} - \zeta^k\frac{\partial g_{kj}}{\partial x^i} \\
&= \frac{\partial \zeta_i}{\partial x^j} + \frac{\partial \zeta_j}{\partial x^i} - 2\zeta_l\Gamma^l_{ij},
\end{aligned}
$$

or

$$\zeta_{i;j} + \zeta_{j;i} = 0. \tag{8.10}$$

Any vector field satisfying this equation is called a *Killing vector*. The problem of finding all of the infinitesimal symmetries of a metric reduces to the problem of finding all Killing vectors. It is interesting to note that the Killing vector equation of the covariant components, expressed in local inertial coordinates at a point, reduces to

$$\zeta_{i,j} + \zeta_{j,i} = 0. \tag{8.11}$$

The form is independent of the signature of the metric. Killing vectors for all metrics satisfy the same set of equations at a point in the coordinates of a local inertial frame. Of course, not every solution of these (local) equations gives rise to a (global) Killing vector. However, if the metric is "flat" and, thus, there exists a global inertial coordinate system, then in such a coordinate system, Eq. (8.11) applies everywhere. Then any solution of these equations is a Killing vector. The covariant Killing vectors of such metrics are all the same, independent of the signature of the metric.

8.3.1   Conserved Momentum

Now that we can characterize the existence of a symmetry of a metric in a way independent of the coordinate system used, we can answer the question, asked in Chapter 7, of how to identify the conserved "momentum" associated with a symmetry. First, note that if a metric is independent of a coordinate $x^{(\beta)}$, then by Eq. (8.9) we have $\zeta^{\alpha}_{(\beta)} = \delta^{\alpha}_{(\beta)}$ as a Killing vector. (By enclosing an index in parentheses we indicate a particular index in contrast to a generic one.) Also, we know that $p_{(\beta)} \equiv mU_{(\beta)}$ is constant along a geodesic—it is conserved. Furthermore, since $\zeta^{\alpha}_{(\beta)} = \delta^{\alpha}_{(\beta)}$, $mU_{(\beta)} = m\zeta^{\alpha}_{(\beta)} U_{\alpha}$. Consider now

$$p_{(\zeta)} \equiv m\zeta^{\alpha} U_{\alpha} = m\zeta_{\alpha} U^{\alpha}, \tag{8.12}$$

where U^{α} is the generalized velocity of a geodesic and ζ_{α} is any Killing vector. $p_{(\zeta)}$ is a scalar, the generalized momentum in the direction of the Killing vector ζ. We can write

$$\frac{dp_{(\zeta)}}{d\tau} = m\frac{d(\zeta^{\alpha} U_{\alpha})}{d\tau} = m\frac{dx^{\sigma}}{d\tau} U^{\alpha}_{;\sigma}\zeta_{\alpha} + U^{\alpha}\zeta_{\alpha;\sigma}\frac{dx^{\sigma}}{d\tau}. \tag{8.13}$$

By use of Eqs. (7.20) and (8.10) we obtain

$$\frac{dp_{(\zeta)}}{d\tau} = U^{\alpha}\zeta_{\alpha;\sigma} U^{\sigma} = 0. \tag{8.14}$$

That is, $p_{(\zeta)}$ is constant along a geodesic. It is conserved.

8.4   Maximally Symmetric Spaces

Any linear combination of Killing vectors with constant coefficients is a Killing vector, and thus Killing vectors, form a linear vector space of some dimension. A natural question to ask is, What is, the maximum possible dimension of this linear vector space—that is, the dimension for a *maximally symmetric space.*[2] Consider, for instance, the two-dimensional sphere, a maximally symmetric space. It is reasonably clear that there exist three linearly independent infinitesimal metric automorphisms corresponding to small rotations about three different axes of the embedding space. Similarly, for the two-dimensional Euclidean space there are three: namely, two infinitesimal translations and a single infinitesimal rotation. But what is the general result?

We have suggested that a knowledge of infinitesimal isometries determines "finite" isometries by successive application. A more surprising result is that a knowledge of the Killing vector field and its first covariant derivative *at a point* determines the Killing vector everywhere. The "commutator" of two covariant derivatives, for a general vector field, is (the covariant form of Eq. (6.43))

$$A_{\nu;\rho;\sigma} - A_{\nu;\sigma;\rho} = -R^{\alpha}{}_{\nu\rho\sigma} A_{\alpha}. \tag{8.15}$$

From this equation, with Eq. (6.44), we find that, for any vector field,

$$A_{\nu;\rho;\sigma} - A_{\nu;\sigma;\rho} + (A_{\sigma;\nu;\rho} - A_{\sigma;\rho;\nu}) + (A_{\rho;\sigma;\nu} - A_{\rho;\nu;\sigma}) = 0.$$

For a Killing vector ζ_i, with the use of Eq. (8.10), this equation gives

$$2\zeta_{i;j;k} - 2\zeta_{i;k;j} - 2\zeta_{k;i;j} = 0.$$

With this result and Eq. (8.15), we have

$$\zeta_{i;j;k} = -R^{p}{}_{kji} \zeta_{p}. \tag{8.16}$$

This equation, with a knowledge of the Killing vector $\zeta_i(X)$ and its derivative $\zeta_{i,j}(X)$ at a *fixed* point X, determines all of the third derivatives and, with the derivatives of Eq. (8.16), all of the higher derivatives. Thus, a power series in $x - X$ can be developed giving $\zeta_i(x)$ for x within the radius of convergence of the series.

The fact, that isometries are completely determined by their local behavior can be used to determine the maximum possible number of independent Killing vectors $\zeta_i^{(n)}$ in N dimensions. There are a maximum of N-independent vectors $\zeta_i^{(n)}(X)$ at a point and $N(N-1)/2$ independent $\zeta_{i;j}^{(n)}(X)$ (recall Eq. (8.10)) for a total maximum of $N(N+1)/2$ independent Killing vectors. A *maximally symmetric* space has

---

[2] The subsequent discussion of maximally symmetric spaces follows somewhat that of Weinberg 1972.

$N(N+1)/2$ independent Killing vectors. For $N = 2$ there are at most three independent Killing vectors in agreement with our previous observation for the two-sphere and two-dimensional Euclidean space.

Consider these three infinitesimal isometries for the two- sphere at the point $X = (\theta, \phi) = (0, \pi/2)$. (One might think that it would be more natural to consider the polar point. However, the metric, in terms of polar coordinates, is singular at that point—some Christoffel symbols diverge—whereas the polar coordinates are locally Euclidean coordinates at X.) Since the Γ vanish at the point, the covariant derivatives reduce to ordinary derivatives, and, thus, the Killing vectors satisfy $\zeta^{(n)}_{i,j} = -\zeta^{(n)}_{j,i}$. To first order in $(x - X)$, we can realize the three independent Killing vectors by

$$\zeta^{(1)}_i(x) = \delta_{i1}$$

$$\zeta^{(2)}_i(x) = \delta_{i2}$$

$$\zeta^{(3)}_i(x) = \phi\delta_{i1} - (\theta - \pi/2)\delta_{i2}.$$

The first two are vectors whose first derivatives vanish at X, whereas for the third, the vector vanishes at the point X. The corresponding isometries are locally two translations and a rotation about the point X. Considering the action of an infinitesimal rotation about the point $X^{(1)} = (\theta, \phi) = (\pi/2, \pi/2)$ on the point X, we see it is the first translation, whereas the infinitesimal rotation about the point $X^{(2)} = (\theta, \phi) = (0, 0)$ is the second. Similarly, we can consider for an N-dimensional space the $N(N+1)/2$ infinitesimal isometries in the vicinity of a point X where the coordinates are chosen to be locally "flat"—the Γ's vanish at X. To first order in $(x - X)$, we can realize the independent Killing vectors by

$$\zeta^{(n)}_i(x) = \delta_{in}, \quad n = 1, \dots N \tag{8.17}$$
$$\zeta^{(lm)}_i(x) = (x^m - X^m)\delta^l_i - (x^l - X^l)\delta^m_i, \quad l, m = 1\dots N, \; l > m. \tag{8.18}$$

(As before, the use of ( ) indicates that the enclosed indices are particular in contrast to the generic, and their value characterizes a particular Killing vector.) The isometries are N "translations," Eq. (8.17), and $N(N-1)/2$ "rotations," Eq. (8.18). Note again that the form of these Killing vectors, expressed in covariant components, does not depend on the signature of the metric. These are the (local) Killing vectors for any maximally symmetric N-dimensional space.

The "translations" of Eq. (8.17) can be characterized in a coordinate-independent manner as

$$\zeta_i^{(n)}(X) \neq 0$$
$$\zeta_{i;j}^{(n)}(X) = 0. \qquad (8.19)$$

The existence of N such Killing vectors implies that the space is *homogeneous* at the point X. There exists an isometry that will carry X into any nearby point.

Similarly, the "rotations" of Eq (8.18) can be chosen so that they are characterized as

$$\zeta_{i;j}^{(lm)}(X) \neq 0$$
$$\zeta_i^{(lm)}(X) = 0. \qquad (8.20)$$

The existence of $N(N-1)/2$ such Killing vectors implies that the space is *isotropic* about the point X.

It is not too difficult to show that if a space is *isotropic* about all points, it is homogeneous at all points. (Just consider effecting the translation of the point X in some arbitrary direction by a rotation centered about a suitable nearby point.) Thus, a space that is isotropic about all points is at any point isotropic and homogeneous and admits the maximum number of Killing vectors—it is *maximally symmetric*. Similarly, a space that is *maximally symmetric* is isotropic and homogeneous, since at each point the $N(N+1)/2$ independent Killing vectors are the N translations and $N(N-1)/2$ rotations, the latter implying isotropy around the point.

It is reasonable that the information about the curvature tensor of a maximally symmetric space can be gained by considering the $N(N-1)/2$ "rotation" Killing vectors, which are those associated with rotation about a point. First, let us note that Eq. (8.16), evaluated at this point for such a rotation Killing vector, provides no useful information. One can see this by considering Eq. (8.16) in a local inertial coordinate system. It merely requires that the second derivatives of the Killing vectors vanish. We must look at higher covariant derivatives.

Note, as is easily shown in a local inertial coordinate system, that a generalization of Eq. (8.15) for rank-2 tensors

$$A_{ij;\,k;\,l} - A_{ij;\,l;\,k} = -R^m_{ikl}\,A_{mj} - R^m_{jkl}\,A_{im}, \tag{8.21}$$

becomes, with $A_{ij} = \zeta^{(mn)}_{i;\,j}$,

$$\zeta^{(mn)}_{i;\,j;\,k;\,l} - \zeta^{(mn)}_{i;\,j;\,l;\,k} = -R^p_{ikl}\,\zeta^{(mn)}_{p;\,j} - R^p_{jkl}\,\zeta^{(mn)}_{i;\,p}. \tag{8.22}$$

Also, the covariant derivative of Eq. (8.16) becomes, with $\zeta_{i;\,j} = \zeta^{(mn)}_{i;\,j}$,

$$\zeta^{(mn)}_{i;\,j;\,k;\,l} = -R^p_{kij;\,l}\,\zeta^{(mn)}_p(X) - R^p_{kij}\,\zeta^{(mn)}_{p;\,l}(X) = -R^p_{kij}\,\zeta^{(mn)}_{p;\,l}(X), \tag{8.23}$$

where $\zeta^{(mn)}_p(X) = 0$ has been used. With use of Eq. (8.23) and the form of $\zeta^{(mn)}_{p;\,l}(X)$ in *local inertial coordinates* given by Eq. (8.18), Eq. (8.22) becomes

$$-R^m_{kij}\,\delta^n_l + R^n_{kij}\,\delta^m_l + R^m_{lij}\,\delta^n_k - R^n_{lij}\,\delta^m_k = -R^m_{ikl}\,\delta^n_j + R^n_{ikl}\,\delta^m_j - R^n_{jkl}\,\delta^m_i + R^m_{jkl}\,\delta^n_i. \tag{8.24}$$

As derived, this equation is valid only at the point X and only in local inertial coordinates. However, since it is assumed that all $N(N-1)/2$ rotations give rise to a Killing vector, this equation is valid for all values of m and n. It follows that Eq. (8.24) is valid, at the point X, in any coordinate system, since it is covariant in form if m and n can take on all values. Furthermore, if, as assumed, the space is isotropic about all points, then Eq. (8.24) is valid at all points. Contracting m with k, using the definition of the Ricci tensor, Eq. (6.47), $R^n_{jil} = -R^n_{lij}$ and $R^m_{mij} = 0$, Eq. (6.45), one obtains

$$(N-1)R^n_{lij} = -R_{il}\,\delta^n_j + R_{jl}\,\delta^n_i. \tag{8.25}$$

This can be rewritten as

$$(N-1)R_{nlij} = -R_{il}\,g_{nj} + R_{jl}\,g_{ni}. \tag{8.26}$$

But $R_{nlij} = -R_{lnij}$, so that we have

$$-R_{il}\,g_{nj} + R_{jl}\,g_{ni} = R_{in}\,g_{lj} - R_{jn}\,g_{li}. \tag{8.27}$$

Now contracting i with n, after "raising" one of the indexes, we find

$$NR_{lj} = R\,g_{lj}. \tag{8.28}$$

Put into Eq. (8.26), this gives

$$R_{nlij} = \frac{R}{N(N-1)}\,(g_{lj}\,g_{ni} - g_{il}\,g_{nj}). \tag{8.29}$$

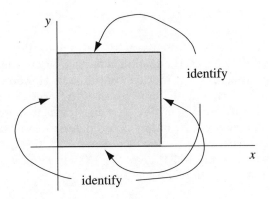

Figure 8.1. Torus.

Eq. (8.29) is a necessary condition that a metric be maximally symmetric. It is not sufficient. Consider, for example, a torus defined by a rectangle in a two-dimensional Euclidean space with opposite sides identified and with the metric $dl^2 = dx^2 + dy^2$.(See Fig 8.1.) Such a space clearly satisfies Eq. (8.28) —it is after all a flat space— but it is not maximally symmetric. It is homogeneous but not isotropic about all points.

It is interesting and useful to determine what information about the metric can be gained by assuming the metric is homogeneous about each point, and, thus, satisfies Eq. (8.19), but not necessarily isotropic. Of course, any result obtained would be valid for isotropic metrics since isotropy about any point implies homogeneity about any point.

Eq. (8.22), with $A_{ij} = \zeta^{(n)}_{i;j}(X) = 0$, becomes

$$\zeta^{(n)}_{i;j;k;l} - \zeta^{(n)}_{i;j;l;k} = 0, \tag{8.30}$$

and the covariant derivative of Eq. (8.16) is

$$\zeta^{(n)}_{i;j;k;l} = - R^{p}_{kij;l} \, \zeta^{(n)}_{p}(X). \tag{8.31}$$

Here we have used Eq. (8.19). Together, Eqs. (8.30) and (8.31) imply

$$-R^{p}_{kij;l} \, \zeta^{(n)}_{p}(X) + R^{p}_{lij;k} \, \zeta^{(n)}_{p}(X) = 0. \tag{8.32}$$

Using Eq. (8.17), valid in a local inertial frame for the N-independent translations, we have

$$-R_{kij;\,l}^{(n)}(X) + R_{lij;\,k}^{(n)}(X) = 0. \tag{8.33}$$

Similar to the above discussion of the implication of assumed isotropy of the metric, this equation is derived to be true only at the point X and only in local inertial coordinates. However, since it is assumed that every one of the N translations gives rise to a Killing vector, this equation is valid for all values of n. It follows that Eq. (8.33) is valid, at point X, in any coordinate system since it is form invariant and true for any value of n. Since $R_{pij}^{p} \equiv 0$ (Eq. (6.45)) and thus $R_{pij;\,l}^{p} \equiv 0$, then Eq. (8.33) gives $R_{lij;\,p}^{p}(X) = 0$. Thus, if the metric is homogeneous everywhere, we see that

$$R_{lij;\,p}^{p}(x) = 0. \tag{8.34}$$

This implies

$$g^{li} R_{lij;\,p}^{p}(x) = (g^{li} R_{lij}^{p}(x))_{;\,p} = R_{j;\,p}^{p}(x) = 0,$$

which by use of Eq. (6.50) gives

$$R_{;\,i} = R_{,\,i} = 0. \tag{8.35}$$

The Ricci scalar is constant—not a suprising result. This of course means that the Ricci scalar in Eq. (8.29) for the Riemann tensor of a maximally symmetric space is a constant.

So, maximally symmetric spaces satisfy Eq. (8.29) with R a constant. Further, all such metrics with the same value of R (and the same signature) are equivalent. That is, if there are two maximally symmetric metrics $g_{kl}(x)$ and $g_{ij}'(x')$ that satisfy Eq. (8.29) with the same value of R, there exist functions $x'^r = x'^r(x)$ such that[3]

$$g_{kl}(x) = g_{ij}'(x') \frac{\partial x'^i}{\partial x^k} \frac{\partial x'^j}{\partial x^l}. \tag{8.36}$$

This is a very useful result. The fact that all maximally symmetric metrics, which satisfy Eq. (8.29) and have the same value for the (constant) Ricci scalar, are equivalent enables us to study these metrics by exhibiting any metric that satisfies Eq. (8.29) and studying the properties of the exhibited metric.

---

[3] See Weinberg (1972) for a proof. In his proof, Weinberg exhibits a power series expansion of the coordinate transformation that satisfies Eq. (8.36).

8.5 Maximally Symmetric Two-Dimensional Riemannian Spaces

We will see that a maximally symmetric three-dimensional Riemannian space, as a three-dimensional subspace of four-dimensional space-time, is important in the study of cosmology. To that end we study such a metric. To simplify the discussion, we first consider maximally symmetric two-dimensional Riemannian spaces. What are the possible maximally symmetric two-dimensional space metrics?

Two possibilities are clear. One metric would be that of the surface of a sphere of radius K, embedded in a three-dimensional Euclidean space, that is, a surface characterized by

$$x^2 + y^2 + z^2 = K^2. \tag{8.37}$$

That this two-dimensional space is maximally symmetric is clear. It admits three independent isometries corresponding to three rotations of the sphere. Furthermore, the metric (induced by the three-dimensional Euclidean metric) is locally Euclidean (spacelike). The circumference C of a "circle" is less then $2\pi r$, where r is its "radius," the geodesic distance of the "circle" from its "center." The space has positive curvature. (See Eq. (6.36).) It is instructive to note that the three-dimensional Euclidean embedding space is flat and, thus, is maximally symmetric. It has six independent Killing vectors. However, the constraint surface, defined by Eq. (8.37), is not invariant under the three translations of the flat space. Thus, we might expect that the induced metric on the sphere might have $6 - 3 = 3$ isometries, which it does.

A second possibility is a flat two-dimensional surface. The three independent isometries are the two translations and one rotation. This is a space of zero curvature for which $C = 2\pi r_0$.

There exists a third type of two-dimensional maximally symmetric space. The space can be visualized as a surface imbedded in a space endowed with a Minkowski metric of one "time" $\bar{t}$ and two space coordinates x, y with the surface characterized by

$$\bar{t}^2 - (x^2 + y^2) = K^2. \tag{8.38}$$

The surface is depicted in Figure 8.2. It is a spacelike surface, which is locally Euclidean. Any two nearby points are spacelike separated with the "distance" defined by that induced by the Minkowski metric. Note that the embedding three-dimensional Minkowski space is flat and thus maximally symmetric, possessing six independent Killing

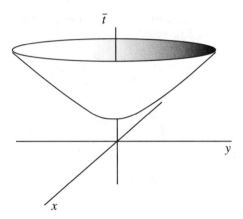

Figure 8.2. Negative curvature surface.

vectors. However, the constraint surface of Eq. (8.38) is not invariant under the three translations of the flat space-time. Thus, we might expect that the induced metric on the sphere might have $6 - 3 = 3$ isometries, and it does. The three independent isometries correspond to the two "Lorentz" boosts in the x and y directions and the rotation in the xy plane. This metric is a maximally symmetric metric of negative curvature in the sense that the circumference C of a "circle" is greater then $2\pi r$. It is instructive to show that this is the case for this surface. Consider a circle defined by the locus of points $x^2 + y^2 = r_0^2$ on the surface. The circumference, using the metric induced on the surface, is simply $2\pi r_0$. The points of this circle are all at minimum distance (i.e., the geodesic distance) r from $x = y = 0$ given by

$$r = \sum (\Delta x^2 + \Delta y^2 - \Delta \bar{t}^2)^{1/2} < r_0 \ . \tag{8.39}$$

$C = 2\pi r_0$ is greater then $2\pi r$.

It is useful to introduce two independent coordinates, one "radial"-like and the other "angle"-like, to characterize points in these spaces and to express the metric in terms of these coordinates. The use of such coordinates will result in similar forms for the three metrics.

8.5.1 Two-Dimensional Space Metric of Positive Curvature

The metric of the imbedding space is

$$ds^2 = dx^2 + dy^2 + dz^2 . \tag{8.40}$$

With the differentials not independent but constrained by Eq. (8.37), this is also the metric induced on the sphere. The differential form of the constraint Eq. (8.37) is

$$2z dz + d(x^2 + y^2) = 0. \tag{8.41}$$

In terms of a "radial" coordinate $r'^2 = x^2 + y^2$, this constraint equation can be written as

$$dz^2 = \frac{r'^2 dr'^2}{K^2 - r'^2}. \tag{8.42}$$

Thus, the invariant distance of points on the sphere, in terms of the polar variables r' and $\theta = \tan^{-1}(x/y)$, is given by

$$ds^2 = \frac{K^2 dr'^2}{K^2 - r'^2} + r'^2 d\theta^2. \tag{8.43}$$

By defining $r' = Kr$, which merely rescales the distance measurement of r', we obtain

$$ds^2 = K^2 \left(\frac{dr^2}{1 - r^2} + r^2 d\theta^2 \right). \tag{8.44}$$

Here r, which is dimensionless, ranges from 0 to 1 and to each value of r and θ there correspond two points on the sphere, one in the "north" hemisphere and one in the "south" hemisphere.

8.5.2   Two-Dimensional Space Metric with Zero Curvature (Flat)

This is easy.

$$ds^2 = K^2 (dx^2 + dy^2) = K^2 (dr^2 + r^2 d\theta). \tag{8.45}$$

8.5.3   Two-Dimensional Space Metric with Negative Curvature

We proceed in a manner similar to that for the positive curvature case. The metric of the embedding space is

$$ds^2 = dx^2 + dy^2 - d\bar{t}^2. \tag{8.46}$$

Again, for the metric on the surface the differentials are not independent but are constrained by Eq. (8.38). Thus, we have

$$-2\bar{t}d\bar{t} + d(x^2 + y^2) = 0. \tag{8.47}$$

In terms of a "radial" coordinate $r'^2 = x^2 + y^2$, this constraint equation reads

$$d\bar{t}^2 = \frac{r'^2\,dr'^2}{K^2 + r'^2}. \tag{8.48}$$

Thus, the invariant distance of points on the surface in terms of the polar variables r' and θ is given by

$$ds^2 = \frac{K^2\,dr'^2}{K^2 + r'^2} + r'^2\,d\theta^2. \tag{8.49}$$

Again defining $r' = Kr$ we obtain

$$d\bar{t}^2 = K^2\left(\frac{dr^2}{1 + r^2} + r^2\,d\theta^2\right). \tag{8.50}$$

Here r, which is dimensionless, ranges from 0 to ∞ and to each value of r and θ there is but one point on the surface.

Combining the results of these three spaces we can write

$$d\bar{t}^2 = K^2\left(\frac{dr^2}{1 - kr^2} + r^2\,d\theta^2\right), \tag{8.51}$$

where $k = \pm 1$ or 0. One may very well ask if these are the most general maximally symmetric metrics. By computing the curvature tensor and noting that Eq. (8.29) is satisfied with R determined in magnitude by K and sign by k, using the uniqueness theorem, we can answer yes. We will return to this after we generalize the results to three space dimensions.

8.6 Maximally Symmetric Three-Dimensional Riemannian Spaces

We could obtain the maximally symmetric three-dimensional Riemannian metric by enlarging the embedding space metrics of Eqs. (8.40) and (8.45) by one space dimension. However, to generalize the results from two to three dimensions, let us merely consider the metric defined by

$$dl^2 = K^2\left(\frac{dr^2}{1 - kr^2} + r^2\,d\theta^2 + r^2\sin^2\theta\,d\phi^2\right), \tag{8.52}$$

where K and k are constants. Here r is a radial-like variable and θ and ϕ are angle variables. The space is rotationally symmetric (isotropic) about $r = 0$ and, for, $k = 0$, is the three-dimensional flat Euclidean space. It is not clear that the space is isotropic about any point for $k \neq 0$. However, computing (say, by use of *Mathematica*), one obtains for the nonvanishing components of the curvature tensor

$$R^r_{\theta r \theta} = -R^r_{\theta \theta r} = R^\phi_{\theta \phi \theta} = -R^\phi_{\theta \theta \phi} = -kr^2$$
$$R^r_{\phi r \phi} = -R^r_{\phi \phi r} = R^\theta_{\phi \theta \phi} = -R^\theta_{\phi \phi \theta} = -kr^2 \sin^2 \theta$$
$$R^\theta_{r \theta r} = -R^\theta_{r r \theta} = R^\phi_{r \phi r} = -R^\phi_{r r \phi} = -\frac{k}{1 - kr^2} \quad , \tag{8.53}$$

and for the Ricci scalar

$$R = -\frac{6k}{K^2}. \tag{8.54}$$

From these results we see that the metrics defined by Eq. (8.52) satisfy Eq. (8.29) if $k = \pm 1$ or 0, a necessary condition when they are maximally symmetric. (As noted, this is not a sufficient condition. The embedding procedure would prove that they are indeed maximally symmetric.) Again by the uniqueness theorem, are the most general maximally symmetric metrics. The coordinates used in Eq. (8.52) treat $r = 0$ as special, but we now know all points are equivalent.

The nonvanishing Christoffel symbols are

$$\Gamma^r_{rr} = \frac{kr}{1 - kr^2}$$
$$\Gamma^r_{\theta\theta} = -r(1 - kr^2)$$
$$\Gamma^r_{\phi\phi} = -r(1 - kr^2)\sin^2 \theta$$
$$\Gamma^\theta_{r\theta} = \Gamma^\theta_{\theta r} = \Gamma^\phi_{r\phi} = \Gamma^\phi_{\phi r} = \frac{1}{r}$$
$$\Gamma^\theta_{\phi\phi} = -\cos\theta\sin\theta$$
$$\Gamma^\phi_{\theta\phi} = \Gamma^\phi_{\phi\theta} = \cot\theta. \tag{8.55}$$

The singular behavior of $\Gamma^\theta_{r\theta} = \Gamma^\theta_{\theta r} = \Gamma^\phi_{r\phi} = \Gamma^\phi_{\phi r}$ at $r = 0$ and of Γ^r_{rr} at $r = 1$ for $k = +1$ reflects the particular coordinate system and is not intrinsic to the metric. Clearly, for any value of k the metric for r small is that of flat Euclidean space expressed in spherical coordinates.

Let us now investigate the properties of these spaces in the large. Since all points are equivalent, to discuss the properties of the most general geodesic we need consider only those passing through $r = 0$. The geodesic equation,

$$\frac{d^2 x^i}{dl^2} + \Gamma^i_{jk} \frac{dx^j}{dl} \frac{dx^k}{dl} = 0, \tag{8.56}$$

for geodesics passing through $r = 0$, reduces to θ and ϕ being constant and to

$$\frac{d^2 r}{dl^2} + \Gamma^r_{rr} \frac{dr}{dl} \frac{dr}{dl} = 0. \tag{8.57}$$

But since

$$\Gamma^r_{rr} \frac{dr}{dl} \frac{dr}{dl} = kr \frac{1}{K^2} g_{rr} \frac{dr}{dl} \frac{dr}{dl} = kr \frac{1}{K^2}, \tag{8.58}$$

Eq. (8.57) has as solutions for $k = +1$, $r = \sin \frac{l}{K}$ and for $k = -1$, $r = \sinh \frac{l}{K}$. For $k = 0$ the solution is of course $r = \frac{l}{K}$. In these solutions l measures the total invariant distance from $r = 0$ along the geodesic. A circular path is characterized by $r = r_0$ and by some fixed value for ϕ. The invariant length circumference of these paths are

$$C = \oint dl = \int_0^{2\pi} K r_0 \, d\theta = 2\pi K r_0. \tag{8.59}$$

These circles have invariant length radii l_0, that is, an invariant distance from $r = 0$, given by

$$l_0 = \begin{cases} K \sin^{-1} r_0, & k = +1 \\ K r_0, & k = 0 \\ K \sinh^{-1} r_0, & k = -1. \end{cases} \tag{8.60}$$

Therefore, the circumference C is

$$C = \begin{cases} 2\pi K \sin \frac{l_0}{K}, & k = +1 \\ 2\pi l_0, & k = 0 \\ 2\pi K \sinh \frac{l_0}{K}, & k = -1. \end{cases} \tag{8.61}$$

What are the properties of these three generically different spaces? In what respects do they differ?

8.6.1 $k = 0$

As noted, the metric is that of three-dimensional flat Euclidian space. The space is infinite in extent. The geodesics are "straight" lines and the usual relationship between circumference and (invariant) radius obtains, $C = 2\pi l_0$.

8.6.2. $k = +1$

The equation for the geodesic, $r = \sin \frac{l}{K}$, tells us that, as we let the geodesic distance l run to $K\pi/2$, the coordinate r reaches its maximum value of unity. This reflects the coordinate singularity in the metric at $r = 1$. However, we can continue the invariant distance to $K\pi$ and the coordinate r returns to the value zero. One might expect that the r coordinate is one which, for fixed values of θ and ϕ, represents more than one point in space, like a radial coordinate for a polar projection of a sphere. Indeed, the metric for $k = +1$ is the induced metric of a three-dimensional sphere imbedded in a four-dimensional Euclidean flat space. As l goes from 0 to $K\pi/2$ and on to $K\pi$, the point on this sphere moves from the "north pole" to the "equator" on to the "south pole." The geodesic continues with l going from $K\pi$ to $K3\pi/2$ to $K2\pi$, with the point going back to the "equator" and returning to the "north pole." So the invariant distance around the space along a geodesic is

$$L = 2\pi K. \tag{8.62}$$

One can say that the space has a radius K. The space is finite but without boundary. From Eq. (8.61) we see that the relationship between the circumference of a circle and the invariant radius is not that of flat space. Rather, the circumference increases more slowly than the invariant radius, just as latitude circles on a two-dimensional sphere. It is a space of positive curvature, as expected.

8.6.3 $k = -1$

The equation of the geodesic $r = \sinh \frac{l}{K}$ tells us that the geodesic distance can increase without bound as does the coordinate r. The space is infinite. And again the relationship between the circumference of a circle and the invariant radius is not that of flat space. But, in contrast to the previous case, the circumference increases more rapidly than the invariant radius. This is a space of negative curvature.

8.7 Maximally Symmetric Four-Dimensional Lorentzian Spaces

Though not having any direct application to the study of cosmology, it is of historical and formal interest to study maximally symmetric space-time metrics, that is, Lorentzian spaces that have ten independent Killing vectors. One such space is of course Minkowski space, which can be characterized as the maximally symmetric four-dimensional pseudo-Riemannian space with signature $(+, -, -, -)$ and with zero Ricci scalar. One might expect that one could obtain another by considering a surface embedded in a five-dimensional pseudo-Riemannian space with signature $(+, +, -, -, -)$ —thus, a surface

$$(x^1)^2 + (x^2)^2 - ((x^3)^2 + (x^4)^2 + (x^5)^2) = (r_1)^2 - (r_2)^2 = a^2 \qquad (8.63)$$

in a space with a metric

$$d\tau^2 = (dx^1)^2 + (dx^2)^2 - ((dx^3)^2 + (dx^4)^2 + (dx^5)^2)$$
$$= dr_1^2 + r_1^2 \, d\theta_1^2 - dr_2^2 - r_2^2 \, d\theta_2^2 - r_2^2 \sin^2(\theta_2) \, d\phi^2 . \qquad (8.64)$$

The induced metric on the surface (after rescaling the r_2 variable) is

$$d\tau^2 = a^2(1 + r^2) \, d\theta_1^2 - \frac{1}{1 + r^2} \, dr^2 - r^2 \, d\theta_2^2 - r^2 \sin^2(\theta_2) \, d\phi^2 . \qquad (8.65)$$

It is clear that the metric, called the anti–de Sitter metric, is Lorentzian. The one timelike variable is θ_1. A computation of the Ricci scalar yields

$$R = -\frac{12}{a^2} , \qquad (8.66)$$

a negative R. A maximally symmetric four-dimensional Lorentzian space with a positive R would complete the set of such metrics. (See Exercise 2, Chapter 9.)

8.8   Exercises

1. The ten independent Poincaré transformations induce an isometry of the Minkowski metric, and thus there are ten independent Killing vectors (the Minkowski metric is maximally symmetric) of the form given by Eqs. (8.17) and (8.18). (Let $X = 0$.) (a) What are the contravariant components of the Killing vectors associated with the four infinitesimal coordinate translations? What are the conserved momentum components associated with these four Killing vectors? (b) What is the Killing vector associated with an infinitesimal Lorentz boost along the x^1 axis? (c) What is the Killing vector associated with an infinitesimal rotation about the x^3 axis? What is the associated conserved (angular) momentum component? (d) By considering the infinitesimal form of the Lorentz transformation (i.e., $\beta \ll 1$), derive the contravariant components of the resulting Killing vector. Do they agree with that obtained in (b)?

2. The Schwarzschild metric is time independent and spherically symmetric. Because in Schwarzschild coordinates the metric is x^0 and ϕ independent, p_0 and p_ϕ are conserved. However, spherical symmetry implies there are two more independent conserved (angular) momentum components. (a) What is the Killing vector associated with the infinitesimal rotation about the x axis? (Consider the polar axis to be the z axis.) Check that this vector satisfies the Killing equation. (b) What is the conserved (angular) momentum component associated with this Killing vector?

3. (a) Derive Eq. (8.65) (b) Show that this metric is maximally symmetric. (c) Obtain Eq. (8.66).

4. Show that the anti–deSitter metric satisfies Einstein's vacuum field equations *with a cosmological constant*. What is the cosmological constant in terms of the constant Ricci scalar?

Chapter 9

Cosmology

9.1 Introduction

We have seen that Einstein's replacement of the Newtonian gravitational force by a change in the geometry of space-time results in changes in physical processes that take place in the vicinity of massive bodies. We now turn our attention to the effect of Einstein's theory on our view of the universe in the large. This study of the large-scale structure and behavior of the universe constitutes the science of cosmology. Prior to Einstein, most astronomers thought the universe to be infinite in extent, Euclidean, and with a uniform and static distribution of stars. Of course, one can see with the naked eye that the distribution of stars is not uniform. What was believed was that, on some large-scale average, the distribution was uniform. Even from a Newtonian standpoint there are difficulties with this view of the universe. First, one expects the gravitational force to be the only significant force acting between stars, and it is rather easily argued that there cannot exist a static distribution of mass points with only the force of gravity acting between them. Second, there is Olbers' paradox, named after the German astronomer Heinrich Olbers (1758–1840). The paradox is between the existence of a rather dark night sky and a static, uniform, and infinite universe.[1] The paradox can be appreciated most simply by realizing that in such a universe if one looked in any direction the line of sight would eventually intercept a star. How bright would such a sky

---

[1] For an extensive discussion of the history and resolution of Olbers' paradox, see Harrison (1987).

be? Consider the light that arrives from a portion of a star's surface that is subtended by a viewing cone of some very small apex angle. This area is proportional to r^2 where r is the distance to the star. The intensity of light received is proportional to $1/r^2$. Thus, the amount of light energy received per unit time within the viewing cone is independent of the distance to the star: the brightness does not depend on the distance to the star. The whole sky would be as bright as a typical star's surface. It is rather obvious that the night sky is not that bright—nor the day sky, for that matter. Note also that such a view of the universe implies a preferred reference frame; namely, one in which the stars are seen, in the average, to be at rest.

Most modern cosmological theories assume the *cosmological principle* according to which all points and directions in the universe are essentially the same. More particularly, the assumption is made that observations made in the universe are spatially homogeneous and isotropic. Of course, the cosmological principle is not exact. There are clumps of matter, stars and galaxies, in the universe. Einstein has taught us that energy-momentum determines geometry. Thus, one implication of the cosmological principle is that the geometry of space, in the large, is homogeneous and isotropic. But to what observers is space homogeneous? And at what "times," for the various observers distributed throughout space, does the universe appear the same? We address these questions in the next section.

The cosmological principle appeals to cosmologists for philosophical and mathematical reasons. For most of the history of modern cosmology, it has been taken as an assumption. Indeed, it is difficult to imagine what form cosmological studies would take if some form of a cosmological principle were not operative. After all, there exists only one universe to study and, if our observation point were not essentially like other points in the universe, what could we learn about the universe in the large by studying it from our viewpoint? There are, however, philosophical and physical questions raised by the special nature of universes that satisfy the cosmological principle, and these have given rise to attempts to derive the homogeneity and isotropy of space.

It may be that the universe does not satisfy the cosmological principle, or at least not exactly—or perhaps not for all time. But it is a unifying idea that makes a good starting point for organizing observational data. The discovery and understanding of violations of the principle would be most interesting.

9.2   The Robertson-Walker Metric

In order to use the cosmological principle we need to formulate mathematically what it says about the space-time metric. Having studied the implications of symmetries on the metric in the last chapter, we are in a good position to do this. To define a time coordinate t, it is useful to use a local scalar property of the universe that is changing, such as the proper energy ρ_0 density of the event point. (This, of course, assumes that ρ is changing.) Here ρ_0 is the energy density as seen by a local inertial observer for whom the local material of the universe is, on the average, not moving, that is, for whom the momentum density T^{io} is zero. We can identify such a time coordinate at all positions in the universe because, by the cosmological principle, all points are equivalent and experience the same history. We will further assume that such points follow space-time geodesics and use as the time coordinate the proper time of such an observer. This we will call "cosmic" time. (The cosmological principle would then imply that the three-dimensional surfaces characterized by fixed cosmic time are maximally symmetric subspaces.)

In terms of the cosmic time t, we write the cosmological invariant interval as

$$d\tau^2 = dt^2 - 2g_{i0}\, dtdx^i - g_{ij}\, dx^i\, dx^j. \qquad (9.1)$$

The coefficient of dt^2 in this metric is unity because t is chosen to be the proper time of a world line characterized by $dx^i = 0$ (i.e., $U^I = 0$). The assumption that such world lines are geodesics puts further restrictions on the metric. The geodesic equation can always be written as (see Eq. (7.23))

$$\frac{dU_\sigma}{d\tau} = \frac{1}{2}\, g_{\zeta\rho,\,\sigma} U^\zeta U^\rho.$$

For a geodesic with $U^i = 0$ and with a metric for which $g_{00} = 1$, this equation becomes

$$\frac{d(g_{oi}U^0)}{dt} = \frac{dg_{oi}}{dt} = 0. \qquad (9.2)$$

The g_{i0} are (cosmic) time independent. Further, the assumed homogeneity and isotropy of space imply the existence of six independent (spacelike) isometries. Under the action of these isometries, cosmic time does not change, which implies, for the associated Killing vectors, $\zeta^0 = 0$. We can argue that there exists a change of coordinates

$$x^i = X^i(x'^i, x'^0)$$
$$t' = x'^0 = x^0 = t,$$
(9.3)

such that $g'_{i0} = 0$. Under this change of coordinates, we find

$$g'_{i0} = \frac{\partial X^k}{\partial x'^i} g_{k0} + \frac{\partial X^k}{\partial x'^i} \frac{\partial X^j}{\partial x'^0} g_{kj} = \frac{\partial X^k}{\partial x'^i} \left(g_{k0} + \frac{\partial X^j}{\partial x'^0} g_{kj} \right).$$
(9.4)

Thus, an X^i that satisfies

$$g_{k0}(X^i, x'^0) + \frac{\partial X^j}{\partial x'^0} g_{kj}(X^i, x'^0) = 0$$
(9.5)

induces a change of coordinates that results in $g'_{i0} = 0$. Eq. (9.5) can be viewed as a coupled set of three ordinary differential equations (for fixed x'^i). This set has a solution with arbitrary initial conditions for X^i at some $x'^0 = x_0^0$, which can be chosen to be x'^i; that is, $X^i(x'^i, x_0'^0) = x'^i$. With these initial conditions, we have $x^i = x'^i$ on the "surface" characterized by $x^0 = x'^0 = x_0^0$. Of course, away from this surface, that is, at $x^0 \neq x_0^0$, generally $X^i(x', x^0) \neq x'^i$.

Thus, the invariant interval of Eq. (9.1) in the new coordinates, for which we drop the primes, becomes

$$d\tau^2 = (dx^0)^2 - g_{ij}(x^i, x^0) dx^i dx^j.$$
(9.6)

The six spacelike isometries still have $\zeta^0 = 0$, and they satisfy (see Eq. (8.9))

$$\frac{\partial g_{ij}(x, x^0)}{\partial x^m} \zeta^m(x, x^0) + g_{il}(x, x^0) \frac{\partial \zeta^l}{\partial x^j} + g_{kj}(x, x^0) \frac{\partial \zeta^k}{\partial x^i} = 0.$$
(9.7)

Only the space-space part of the metric is involved. The metric $g_{ij}(x, x^0)$ is maximally symmetric for each value of x^0. Eq. (8.52) in the previous chapter implies that space-space coordinates exist for which the space-space invariant interval has the form

$$dl^2 = K^2(x^0) \left(\frac{dr^2}{1 - kr^2} + r^2 d\theta^2 + r^2 \sin^2\theta d\phi^2 \right).$$
(9.8)

The constant k cannot change with time since it has discrete values and thus cannot change in a continuous manner. Finally, we have for the general metric satisfying the cosmological principle,

$$d\tau^2 = dt^2 - K^2(t)\left(\frac{dr^2}{1 - kr^2} + r^2\,d\theta^2 + r^2 \sin^2\theta d\phi^2\right), \qquad (9.9)$$

with $k = 0, \pm 1$. This metric is known as the *Robertson-Walker metric*. The (large-scale average) metric history of a universe satisfying the cosmological principle is contained in the function $K(t)$ and the constant k. Knowledge of $K(t)$ and the constant k must be gained by observation and theoretical interpretation. Note that no dynamics, such as Einstein's general relativity equations, have been used in determining the form of Eq. (9.9). Such dynamics can tell us about the possible time dependence of $K(t)$, its relation to k, and the relation to other physical entities such as the mass density of the universe. Before moving on to this, we will first study the kinematics of the Robertson-Walker metric.

9.3   Kinematics of the Robertson-Walker Metric

9.3.1   Proper Distance

Recall that a "typical" galaxy (or material) moves along a geodesic characterized by fixed r, θ, and ϕ with the cosmic time t being the proper time of the galaxy (or material). These coordinates are referred to as *co-moving coordinates*. We assume $r = 0$ is our galaxy's position. This is just a choice of coordinate system, not an observation that our galaxy is special—recall the cosmological principle. We would like to know how far apart, at a given cosmic time t, are two galaxies positioned at $r = 0$ and $r = r_1$. But first we will answer the question of how far apart, dl, two galaxies are at r and $r + dr$ (same angle coordinates). We define this distance as the cosmic time it takes light to travel from r to $r + dr$. Since we have for the world line of a light particle $d\tau^2 = 0$, we find

$$dl = K(t)\frac{dr}{(1 - kr^2)^{1/2}}, \qquad (9.10)$$

which changes with time as $K(t)$ changes. The *proper distance* d between the galaxies at $r = 0$ and $r = r_1$ at a fixed cosmic time t is then defined by

$$d(t, r_1) = K(t)\int_0^{r_1} \frac{dr}{(1 - kr^2)^{1/2}}. \qquad (9.11)$$

As measured by proper distance, typical galaxies move apart or together as $K(t)$ increases or decreases. The quantity d is not the time it takes light to travel from a typical galaxy at r_1 to our galaxy. After all, d is a distance measurement at fixed cosmic time. But a photon which reaches our galaxy travels "radially," and, thus, satisfies

$$d\tau^2 = dt^2 - \frac{K^2(t)\,dr^2}{1 - kr^2} = 0.$$

From this we see that the cosmic time t_1 at which light was emitted at r_1 to arrive at our galaxy at the present time t_0 is determined by the relation

$$\int_{t_1}^{t_0} \frac{dt}{K(t)} = \int_0^{r_1} \frac{dr}{(1 - kr^2)^{1/2}}. \tag{9.12}$$

Another useful "proper" distance is that which is perpendicular to a radial distance. Imagine that we observe an object, say a galaxy, that suspends an angle $\delta\theta$ (in some direction). What is the proper length of that object? The light signals we observe from the two ends of the object, separated by a small angle $\delta\theta$, are emitted at the same cosmic time t_1 and at the same r value, say r_1. From the operational definition of proper distance given above, the proper (perpendicular) distance $l_\perp$ between these positions at the cosmic time the light was emitted is given by

$$l_\perp = K(t_1)\,r_1\,\delta\theta. \tag{9.13}$$

Here $K(t_1)$ and r_1 are related by Eq. (9.12).

These considerations, particularly Eq. (9.12), can be used to illustrate the concept of horizons, that is, the limits of our view of the universe. In the next two sections we will consider two types of horizons, called *particle horizons* and *event horizons* by Rindler (1956).

9.3.2 Particle Horizons

Our particle horizon $r_{PH}(t_0)$ is defined to be the largest value of r from which we could have received a light signal up to the present. More accurately, it is the position $r_{PH}(t_0)$ whose light signal, emitted at the earliest possible time, is just now reaching us. Of course, its value depends on the time dependence of $K(t)$. Eq. (9.12) gives

$$\int_{0,\,-\infty}^{t_0} \frac{dt}{K(t)} = \int_0^{r_{PH}(t_0)} \frac{dr}{(1 - kr^2)^{1/2}}. \tag{9.14}$$

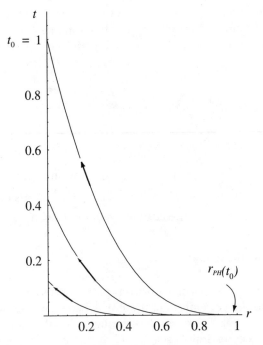

Figure 9.1. World lines of light emitted at the big bang.

The lower limit of the time integral is 0 if $K(t)$ goes to zero in the past at the cosmic time chosen to be zero (the "big bang" event), whereas, if such an event did not occur, the limit is $-\infty$. A particle horizon will exist if

$$\int_{0,\,-\infty}^{t_0} \frac{dt}{K(t)} < \infty. \tag{9.15}$$

For the big bang case, this inequality is satisfied if $K(t)$ vanishes more slowly than $t^{1-\epsilon}$.

As an illustration of the particle horizon, Figure 9.1 shows the world lines of light particles that were emitted toward us at time zero at various r positions. (Remember r is a co-moving coordinate.) t_0 is chosen to be equal to one. The lines are calculated for the $k = 0$ universe and with the dynamic assumption that the universe can be described as a dust field universe with zero pressure all the way back to time zero. (See Sec. 9.4.) The three world lines in the figure reach $t = 0$ at $r = 1$, $3/4$, and $1/2$. For this case $K(t)$ vanishes as $t^{2/3}$. We see that any light emitted at $r > r_{PH}(t_0)$ at time $t = 0$ has not yet reached us, whereas light emitted from $r < r_{PH}(t_0)$ has gone past us.

The integral on the right side of Eq. (9.14) can be evaluated to yield

$$\int_{0,-\infty}^{t_0} \frac{dt}{K(t)} = \begin{cases} \sin^{-1} r_{PH}(t_0), & k = +1 \\ r_{PH}(t_0), & k = 0 \\ \sinh^{-1} r_{PH}(t_0), & k = -1. \end{cases} \tag{9.16}$$

For $k = +1$, if $\int_{0,-\infty}^{t_0} dt/K(t) > 2\pi$, the horizon photon will have passed us at least once before. We will return to the discussion of particle horizons after we study the cosmic dynamics of general relativity.

9.3.3   Event Horizons

Just as there may be particle horizons past which we have observed no events up to the present time, there may be events that we can never see. From Eq. (9.12) we see that the light signal of an event, occurring at r_1 at cosmic time t_1, reaches us at cosmic time t determined by

$$\int_{t_1}^{t} \frac{dt}{K(t)} = \int_0^{r_1} \frac{dr}{(1 - kr^2)^{1/2}}. \tag{9.17}$$

Thus, there exists an event horizon $r_{EH}(t_1)$ for an event occurring at cosmic time t_1 if

$$\int_{t_1}^{t_{max}} \frac{dt}{K(t)} < \infty. \tag{9.18}$$

The event horizon $r_{EH}(t_1)$ is determined by

$$\int_{t_1}^{t_{max}} \frac{dt}{K(t)} = \int_0^{r_{EH}} \frac{dr}{(1 - kr^2)^{1/2}}, \tag{9.19}$$

where t_{max} is finite if at some time in the future $K(t) \to 0$, or infinity if not. As before, this can be written as

$$\int_{t_1}^{t_{max}} \frac{dt}{K(t)} = \begin{cases} \sin^{-1} r_{EH}(t_1), & k = +1 \\ r_{EH}(t_1), & k = 0 \\ \sinh^{-1} r_{EH}(t_1), & k = -1. \end{cases} \tag{9.20}$$

One should appreciate that such event horizons have no observational consequence but might be of philosophical interest.

9.3.4 Cosmological Redshift: Hubble's Constant

The time dependence of K has observational consequences, and thus observational astronomy contributes to our knowledge of K. The best and most direct information is obtained from the shift in the frequency of light coming from distant galaxies. We can calculate this frequency by considering two successive crests of a light wave that travels from a galaxy at r_1 and is received by us (at $r = 0$). Each crest satisfies the equation

$$dt^2 - K^2(t) \frac{dr^2}{(1 - kr^2)} = 0. \tag{9.21}$$

If a crest is emitted at t_1 from r_1, it is received at the origin at t_0, given by

$$\int_{t_1}^{t_0} \frac{dt}{K(t)} = - \int_{r_1}^{0} dr (1 - kr^2)^{-1/2}. \tag{9.22}$$

If the next crest is emitted at time $t_1 + \delta t_1$ (from r_1), it is received at the origin at a time $t_0 + \delta t_0$ given by

$$\int_{t_1 + \delta t_1}^{t_0 + \delta t_0} \frac{dt}{K(t)} = \int_{r_1}^{0} dr (1 - kr^2)^{-1/2}. \tag{9.23}$$

Thus, we have

$$\int_{t_1 + \delta t_1}^{t_0 + \delta t_0} \frac{dt}{K(t)} - \int_{t_1}^{t_0} \frac{dt}{K(t)} = \int_{t_0}^{t_0 + \delta t_0} \frac{dt}{K(t)} - \int_{t_1}^{t_1 + \delta t_1} \frac{dt}{K(t)} = 0. \tag{9.24}$$

Since $K(t)$ does not change significantly during the period of a typical light wave, we find that

$$\frac{\delta t_0}{K(t_0)} = \frac{\delta t_1}{K(t_1)}. \tag{9.25}$$

Of course $t_0 > t_1$. Since cosmic time is the proper time of typical galaxies, δt_0 is the period of the light received, and δt_1 is the period emitted. The frequencies are related by

$$\frac{\nu_0}{\nu_1} = \frac{K(t_1)}{K(t_0)}, \tag{9.26}$$

where ν_0 is the frequency received and ν_1 is the frequency emitted. For the corresponding wavelengths, the relation is

$$\frac{\lambda_0}{\lambda_1} = \frac{K(t_0)}{K(t_1)}. \tag{9.27}$$

If $K(t_1) > K(t_0)$ then $\nu_0 > \nu_1$ and a "blue" shift results, whereas if $K(t_1) < K(t_0)$ then $\nu_0 < \nu_1$ and a red shift obtains. To be able to measure this shift, we must know the frequency emitted, which is not possible for a single frequency. However, by identifying sets of frequencies as those of a particular emission spectrum, thus determining sets of ν_i's, the ratio $K(t_0)/K(t_1)$ can be determined. Observations from distant galaxies give a redshift indicating $K(t_1) < K(t_0)$. The shift is usually expressed in terms of the *redshift parameter* z defined as the fractional *increase* in the wavelength

$$z = \frac{\lambda_0 - \lambda_1}{\lambda_1}, \tag{9.28}$$

which, by Eq. (9.26), is

$$z = \frac{K(t_0)}{K(t_1)} - 1. \tag{9.29}$$

This redshift is sometimes mistakenly referred to as a Doppler shift. It does have a Doppler component. For emission from nearby galaxies, for which r_1 and $t_0 - t_i$ are small, we have

$$z \approx \frac{\dot{K}(t_0)(t_0 - t_1)}{K(t_0)} \approx r_1 \dot{K}(t_0) \approx \dot{d}(t_0, r_1). \tag{9.30}$$

The last equality follows from the definition of proper distance Eq. (9.11). This z is the result for the Doppler shift for low relative velocity $\dot{d}(t_0, r_1)$. However, for light from far galaxies, for what cosmic time would one calculate the "relative" velocity? Rather, the redshift is due to the expansion (or contraction) of space itself during the transit time of the light.

Since, for nearby galaxies, the proper distance $d(t_0, r_1)$ is the time of passage $(t_0 - t_1)$, we can use Eq. (9.30) to obtain a proper distance d redshift z relationship:

$$z \approx \frac{\dot{K}(t_0)(t_0 - t_1)}{K(t_0)} = \frac{\dot{K}(t_0)}{K(t_0)} d = H_0 d. \tag{9.31}$$

The constant of proportionality H_0,

$$H_0 = \frac{\dot{K}(t_0)}{K(t_0)}, \tag{9.32}$$

is called *Hubble's constant* after the astronomer Edwin Hubble. It was he who first announced, in 1929, that observations "establish a roughly linear relation between velocities and distances among nebulae" (Hubble 1929). The most recent measurement of this constant is presented in an article appropriately titled "Final Results from the

Hubble Space Telescope Key Project to Measure the Hubble Constant" (Freedman et al. 2001). The result is a value of about 70 km s^{-1} Mpc^{-1} or a Hubble time H_0^{-1} of 1.3×10^{25} m or 13.8×10^9 years.[2] One should appreciate that the Hubble time, H_0^{-1}, is a natural timescale for the *present* universe. Similarly, $K(t_0)$ is a natural size scale. We will at times find it useful to define a dimensionless time variable $\bar{t} = H_0 t$ and a dimensionless size variable $\bar{K}(\bar{t}) = K(t)/K(t_0)$ and write equations in terms of these variables.

The measurement of the redshift parameter of close galaxies gives us some information about $K(t)$, namely, $\dot{K}(t_0)/K(t_0)$. But, if the cosmologist is to obtain further knowledge of the time dependence of $K(t)$, measurements of redshifts of more distance galaxies have to be related to the time dependence. Thus, it is necessary to define some measurable "distance" that depends on $K(t)$. The most useful distance is the *luminosity distance*.

9.3.5   Luminosity Distance

Suppose that astronomers have identified a standard type of galaxy (or star) and have data that indicates the total power P that it radiates. Then an observation of the luminosity (the power received per unit area) of such a galaxy gives information about the distance of the galaxy. If the space is Minkowskian and if the distant galaxy is at rest with respect to the observer, the luminosity L is given by

$$L = \frac{P}{4\pi r^2}. \tag{9.33}$$

This is just the power divided by the area of a sphere at a distance r. Imagine that the galaxy is positioned at the center of a sphere of radius r and is radiating uniformly in all directions. How is this relation modified in a Robertson-Walker metric? What is the relation between the luminosity and the position of the standard typical galaxy at r? Two effects modify the relation. First, the photons undergo a cosmological redshift, and second, the rate at which the photons are received is not the same as the rate at which they were emitted. We choose a coordinate system with the radiating galaxy at $r = 0$ radiating at cosmic time $t = t_1$, and our receiving galaxy is at $r = r_1$ and receiving at $t = t_0$. By Eq. (9.26), a photon emitted with energy $E_e = h\nu_e$ is received with energy $E_r = h\nu_e K(t_1)/K(t_0)$. Similarly, by Eq.

[2] The Mpc, or megaparsec, is a unit of distance. 1 Mpc $= 3.26 \times 10^6$ light years $= 3.08 \times 10^{22}$ m.

(9.25), if the time between emission of photons is δt_e, then the time between reception at the "sphere" at r_1 is $\delta t_r = \delta t_e K(t_0)/K(t_1)$. From these it follows that the power received on the sphere at r_1 at time t_0 is $PK^2(t_1)/K^2(t_0)$. The area of the sphere at r_1 at cosmic time t_0 is $4\pi K^2(t_0)r_1^2$, and thus the luminosity at r_1 at t_0 is

$$L = \frac{P}{4\pi K^2(t_0)r_1^2} \frac{K^2(t_1)}{K^2(t_0)} = P\frac{K^2(t_1)}{4\pi K^4(t_0)r_1^2}. \qquad (9.34)$$

If we define the *luminosity distance* d_{L_1} for a typical galaxy at r_1 to be

$$d_{L_1} = \frac{K^2(t_0)r_1}{K(t_1)} = \frac{K(t_0)r_1}{\bar{K}(\bar{t}_1)}, \qquad (9.35)$$

then

$$L = \frac{P}{4\pi d_{L_1}^2}. \qquad (9.36)$$

By measuring the luminosity of a standard galaxy, that is, one whose power is known, the astronomer can determine the luminosity distance. Note that $d_{L_1} \to d_1$ for small r_1.

9.3.6 Cosmological Redshift: Deceleration Parameter

Measurement of the redshifts and luminosities for more distant galaxies gives the astronomer more information about the time dependence of K than that given by H_0. So, we might consider a power series expansion in $t - t_0$ of K as

$$K(t) = K(t_0)\left[1 + H_0(t - t_0) - \frac{1}{2}q_0 H_0^2(t - t_0)^2 + ...\right]. \qquad (9.37)$$

or, equivalently,

$$\bar{K}(\bar{t}) = 1 + (\bar{t} - \bar{t}_0) - \frac{1}{2}q_0(\bar{t} - \bar{t}_0)^2 + ... \qquad (9.38)$$

(Recall that we have defined $\bar{K}(\bar{t}) = K(t)/K(t_0)$) and $\bar{t} = H_0 t$.) Hubble's constant, H_0, has already been introduced, and q_0 is known as the *deceleration parameter*.

$$q_0 \equiv -\ddot{K}(t_0)\frac{K(t_0)}{\dot{K}^2(t_0)} = -\ddot{\bar{K}}(\bar{t}). \qquad (9.39)$$

What does one measure to determine q_0? One can measure (hopefully) luminosity distances and redshifts for standard galaxies. What is needed is a luminosity distance-redshift relationship that involves both H_0 and q_0 to replace the proper distance-redshift relationship, Eq. (9.31). One can obtain this relationship by first expanding Eq. (9.29) in a power series in $t - t_0$ and inverting it to obtain

$$t_0 - t_1 = \frac{1}{H_0}\left[z - \left(1 + \frac{q_0}{2}\right)z^2\ldots\right]. \qquad (9.40)$$

Using Eq. (9.12) to expand r_1 to order $(t_0 - t_1)^2$, we obtain

$$r_1 = \frac{1}{K(t_0)}\left[t_0 - t_1 + \frac{1}{2}H_0(t_0 - t_1)^2\ldots\right]. \qquad (9.41)$$

After expanding the time dependence of Eq. (9.35) and substituting Eqs. (9.41) and then (9.40) into that, one obtains the luminosity distance-redshift relationship, for small z,

$$d_{L_1} = H_0^{-1}\left[z + \frac{1}{2}(1 - q_0)z^2\ldots\right]. \qquad (9.42)$$

By measuring luminosity distance and redshifts of galaxies, the astronomer can determine H_0 and q_0. The measurement of luminosity distance is, as one can imagine, fraught with error. How "standard" are standard galaxies? Do standard galaxies age? That is, do they change so that a nearby standard galaxy differs from a distant standard galaxy that is, after all, being viewed at an earlier cosmic time?

Of course, Eq. (9.42) is meaningful only for $z < 1$ and is merely a parameterization of the luminosity distance-redshift relationship. After studying the dynamics of the Robertson-Walker metric, we will obtain a d_{L_1} versus z relationship valid for larger z under particular dynamical assumptions.

9.4   Dynamics of the Robertson-Walker Metric

The dynamics of the Robertson-Walker metric are governed by the Einstein field equation

$$G_{\alpha\beta} + \lambda g_{\alpha\beta} = R_{\alpha\beta} - \frac{1}{2}Rg_{\alpha\beta} + \lambda g_{\alpha\beta} = -8\pi\tilde{G}T_{\alpha\beta}. \qquad (9.43)$$

We allow for the possibility of a nonvanishing cosmological constant λ.

Note that the cosmological principle applies to the energy-momentum tensor $T_{\alpha\beta}$. That is, it should be form invariant under the six spacelike isometry transformations for fixed t. In the coordinates of the R-W metric, these isometry transformations transform only the space-space part of the metric g_{ij}. But the space-space part of the energy-momentum T_{ij} transforms like the g_{ij} so that the form invariance of the T_{ij} would be assured if

$$T_{ij} = -f_1(t) g_{ij}, \tag{9.44}$$

where $f_1(t)$ is some function of t. Similarly, under these isometry transformations, T_{00} transforms the same as g_{00}, that is, it does not change, and we might expect that

$$T_{00} = f_2(t), \tag{9.45}$$

where $f_2(t)$ is some second function of t. Since the components g_{i0} vanish and are not effected by the six isometries, if we assume that T_{i0} vanish and that Eqs. (9.44) and (9.45) are valid, the form invariance of $T_{\alpha\beta}$ would be assured. We can write such a tensor as

$$T_{\alpha\beta} = (f_1 + f_2) U_\alpha U_\beta - f_1 g_{\alpha\beta}, \tag{9.46}$$

where U_α is the four-velocity of a typical galaxy which in Robertson-Walker coordinates becomes $U_\alpha = \delta_{\alpha 0}$. This is the energy-momentum tensor of a perfect fluid with $f_1 = p$, the pressure, and $f_2 = \rho$, the proper energy density. (See Sec. 4.4.2.) We expect the energy-momentum tensor of a universe satisfying the cosmological principle to be that of a perfect fluid, so that

$$T_{\alpha\beta} = (p + \rho) U_\alpha U_\beta - p g_{\alpha\beta}. \tag{9.47}$$

Indeed, if one computes the Einstein tensor $G_{\alpha\beta}$ (using, say, *Mathematica*) for the Robertson-Walker metric, one finds that

$$\begin{aligned} G_{00} &= F_2(t) \\ G_{ij} &= F_1(t) g_{ij} \\ G_{i0} &= 0, \end{aligned} \tag{9.48}$$

with

$$F_2 = -\frac{3(k + \dot{K}(t)^2)}{K(t)^2}$$

$$F_1 = -\frac{k + \dot{K}(t)^2 + 2K(t)\ddot{K}(t)}{K(t)^2}. \tag{9.49}$$

Eq. (9.48), along with Einstein's field equations, implies that the energy-momentum tensor of a universe satisfying the cosmological

principle *must* be that of a perfect fluid (Eq. (9.47)). With this result
and Eq. (9.49), the Einstein field equations can be written as

$$\frac{3(k + \dot{K}(t)^2)}{K(t)^2} - \lambda = 8\pi\tilde{G}\rho(t) \tag{9.50}$$

and

$$-\frac{k + \dot{K}(t)^2 + 2K(t)\ddot{K}(t)}{K(t)^2} + \lambda = 8\pi\tilde{G}p(t). \tag{9.51}$$

In addition to these equations relating the $\rho(t)$ and $p(t)$ to the metric,
the motion of the fluid is governed by the conservation of the energy-
momentum tensor $T_{;\beta}^{\alpha\beta} = 0$. For the case of an ideal fluid, these conservation equations become

$$T_{;\beta}^{\alpha\beta} = \left((p + \rho)U^\alpha U^\beta\right)_{;\beta} - p_{,\beta}g^{\alpha\beta}$$

$$= (p + \rho)_{,\beta}U^\alpha U^\beta + (p + \rho)\left[(U^\alpha U^\beta)_{,\beta} + \Gamma_{\sigma\beta}^\alpha U^\sigma U^\beta + \Gamma_{\sigma\beta}^\beta U^\sigma U^\alpha\right]$$

$$- p_{,\beta}g^{\alpha\beta} = 0. \tag{9.52}$$

Here we have used $g_{;\beta}^{\alpha\beta} = 0$ and the scalar character of p and ρ. For the
Robertson-Walker metric, $U^\alpha = \delta^{\alpha 0}$, and thus Eq. (9.52) becomes

$$T_{;\beta}^{\alpha\beta} = (\dot{p} + \dot{\rho})\delta^{\alpha 0} + (p + \rho)(\Gamma_{00}^\alpha + \Gamma_{0\beta}^\beta\delta^{\alpha 0}) - \dot{p}\delta^{\alpha 0} = 0. \tag{9.53}$$

Furthermore, for the Robertson-Walker metric, we find that

$$\Gamma_{0\beta}^\beta = \frac{3\dot{K}}{K}$$

$$\Gamma_{00}^\alpha = 0. \tag{9.54}$$

Putting these into Eq. (9.53), we obtain

$$T_{;\beta}^{\alpha\beta} = (\dot{p} + \dot{\rho})\delta^{\alpha 0} + (p + \rho)\frac{3\dot{K}}{K}\delta^{\alpha 0} - \dot{p}\delta^{\alpha 0} = 0. \tag{9.55}$$

The space components of this equation are trivially satisfied, while
the time component is

$$\dot{\rho} + 3(p + \rho)\frac{\dot{K}}{K} = 0$$

$$\dot{\rho}(t) + 3(p + \rho)\frac{\dot{K}}{K} = 0, \tag{9.56}$$

which can also be written as

$$\frac{d(p+\rho)K^3}{dt} = \dot{p}K^3$$

$$\frac{d(p+\rho)\bar{K}^3}{d\bar{t}} = \dot{p}\bar{K}^3. \tag{9.57}$$

With the time dependence of ρ and p viewed as given implicitly by their $K(t)$ dependence, Eq. (9.57) can finally be expressed as

$$\frac{d(\rho K^3)}{dK} = -3pK^2$$

$$\frac{d(\rho \bar{K}^3)}{d\bar{K}} = -3p\bar{K}^2. \tag{9.58}$$

This is a useful form of the conservation of energy-momentum equation. Given the equation of state $p = p(\rho)$, Eq. (9.58) presumably can be integrated to give $\rho(K)$, with one integration constant $\rho(K_0)$. This result can be inserted into Eq. (9.50) to be solved for $K(t)$, again with one integration constant $K(t_0)$. We will return to these considerations later.

Note that if the two first-order differential equations, Eqs. (9.57) and (9.50), are satisfied, then the second-order equation, Eq. (9.51), is automatically satisfied. This is not surprising, since the local conservation of the energy-momentum tensor is implied by the Bianchi identities. Nevertheless, the second-order equation obtained by eliminating $\dot{K}(t)$ from Eq. (9.51), with the aid of Eq. (9.50), proves useful:

$$\frac{\ddot{K}(t)}{K(t)} = -\frac{4\pi\tilde{G}}{3}\left(\rho(t) + 3p(t)\right) + \frac{\lambda}{3}. \tag{9.59}$$

Furthermore, we can write the two controlling dynamical equations, Eqs. (9.50) and (9.59), using the time variable $\bar{t}$ and the "scale variable" $\bar{K}$. Note $\bar{K}(\bar{t}_0) = 1$ and $\dot{\bar{K}}(\bar{t}_0)/\bar{K}(\bar{t}_0) = 1$. In terms of these variables, Eq. (9.50) becomes

$$\left(\frac{\dot{\bar{K}}(\bar{t})}{\bar{K}(\bar{t})}\right)^2 = \Omega_{\rho 0}\frac{\rho(\bar{t})}{\rho_0} + \Omega_k \bar{K}(\bar{t})^{-2} + \Omega_\lambda, \tag{9.60}$$

and Eq. (9.59)

$$\frac{\ddot{\bar{K}}(\bar{t})}{\bar{K}(\bar{t})} = -\frac{1}{2}\Omega_{\rho 0}\frac{\rho(\bar{t}) + 3p(\bar{t})}{\rho_0} + \Omega_\lambda. \tag{9.61}$$

Here we have defined the quantities

$$\Omega_{\rho_0} = \frac{8\pi\tilde{G}}{3H_0^2}\,\rho(t_0)$$

$$\Omega_k = -\frac{k}{H_0^2\,K(t_0)^2}$$

$$\Omega_\lambda = \frac{\lambda}{3H_0^2}\,.$$
(9.62)

These are related as

$$1 = \frac{8\pi\tilde{G}}{3H_0^2}\,\rho(t_0) - \frac{k}{H_0^2\,K(t_0)^2} + \frac{\lambda}{3H_0^2}$$

$$= \Omega_{\rho_0} + \Omega_k + \Omega_\lambda\,.$$
(9.63)

The $\Omega_{\rho_0}, \Omega_k$ and Ω_λ can be viewed as the (dimensionless) energy density contribution to H_0 from matter, curvature, and the cosmological constant, respectively. Note that the value of Ω_k, if not zero, determines the value of $H_0 K(t_0)$.

9.4.1 Critical Density

Astronomers can determine Hubble's constant H_0, the deceleration parameter q_0, by measuring the luminosity distance and the redshifts of galaxies, and hopefully the present average density ρ_0. How does the dynamics relate these "kinematic" measurements to the constant k? From Eq. (9.63) we see that k is negative or positive as $1 - \Omega_{\rho_0} - \Omega_\lambda$ is greater or less than zero or as $\rho(t_0)$ is greater or less than a *critical density*,

$$\rho_c = \frac{3}{8\pi\tilde{G}}\,H_0^2 - \frac{\lambda}{8\pi\tilde{G}}\,.$$
(9.64)

If $\lambda = 0, H_0^2$ and $\rho(t_0)$ determine the value of k.

Now consider the second-order dynamical Eq. (9.61) evaluated at the present time $\bar{t}_0$ for which $p(\bar{t}_0) \approx 0$. Using the definition of q_0, Eq. (9.39), we find from Eq. (9.61) that

$$q_0 = \frac{\Omega_{\rho_0}}{2} - \Omega_\lambda\,.$$
(9.65)

The deceleration parameter is positive if $\lambda = 0$ and negative only if $\Omega_\lambda > \Omega_{\rho_0}/2$.

9.4.2 Cosmological Redshift: Distant Objects

As noted before, the luminosity distance versus redshift relation, Eq. (9.42), is merely a power-series development of the luminosity distance in terms of the redshift expected to be accurate only for $z < 1$ at best. But astrophysical objects, quasars, have been observed at apparent redshifts up to 3 and the microwave background, discussed in the next section, has a redshift of about 10^3. Thus, having a distance-redshift relation for $z > 1$, or even one that is more exact for $z \approx .5$, is useful. Such a relation depends of course on the equation of state governing the dynamics of $\bar{K}(\bar{t})$. A universe of "dust" is a good approximation for the present epoch. One expects that the dust approximation should be valid back past the cosmic time at which we are observing quasars, but surely not back to the decoupling time of the microwave background radiation (see next section).

To obtain a relation applicable to $z > 1$, back during the time for which the dust approximation is valid, we return to Eq. (9.12) and change the time integral to a $\bar{K}$ integral. Thus, Eq. (9.12) becomes

$$\frac{1}{K(t_0)H_0} \int_{\bar{K}(\bar{t}_1)}^{\bar{K}(\bar{t}_0)} \frac{d\bar{K}}{\dot{\bar{K}}(\bar{t})\bar{K}} = \int_0^{r_1} \frac{dr}{(1-kr^2)^{1/2}}. \tag{9.66}$$

After evaluating the right-hand integral, we can write this as

$$r_1 = \begin{cases} \sin\left(\dfrac{1}{K(t_0)H_0} \displaystyle\int_{\bar{K}(\bar{t}_1)}^{\bar{K}(\bar{t}_0)} \dfrac{d\bar{K}}{\dot{\bar{K}}(\bar{t})\bar{K}} \right), & k = +1 \\[3mm] \dfrac{1}{K(t_0)H_0} \displaystyle\int_{\bar{K}(\bar{t}_1)}^{\bar{K}(\bar{t}_0)} \dfrac{d\bar{K}}{\dot{\bar{K}}(\bar{t})\bar{K}}, & k = 0 \\[3mm] \sinh\left(\dfrac{1}{K(t_0)H_0} \displaystyle\int_{\bar{K}(\bar{t}_1)}^{\bar{K}(\bar{t}_0)} \dfrac{d\bar{K}}{\dot{\bar{K}}(\bar{t})\bar{K}} \right), & k = -1. \end{cases} \tag{9.67}$$

For a dust-filled universe, we have $p \approx 0$. This fact and Eq. (9.58) give

$$\rho = \rho_0 \bar{K}^{-3}. \tag{9.68}$$

Thus, for a dust-filled universe, Eq. (9.60) becomes

$$\dot{\bar{K}}^2 = \Omega_{\rho_0} \bar{K}^{-1} + \Omega_k + \Omega_\lambda \bar{K}^2, \tag{9.69}$$

which, if used in Eq. (9.67), gives

$$
r_1 = \begin{cases} \sin\left(\dfrac{1}{K(t_0)H_0} \displaystyle\int_{\bar{K}(\bar{t}_1)}^{\bar{K}(\bar{t}_0)} \dfrac{d\bar{K}}{\left(\Omega_{\rho o}\bar{K} + \Omega_x \bar{K}^2 + \Omega_\lambda \bar{K}^4\right)^{1/2}}\right), & k = +1 \\[20pt] \dfrac{1}{K(t_0)H_0} \displaystyle\int_{\bar{K}(\bar{t}_1)}^{\bar{K}(\bar{t}_0)} \dfrac{d\bar{K}}{\left(\Omega_{\rho o}\bar{K} + \Omega_\lambda \bar{K}^4\right)^{1/2}}, & k = 0 \\[20pt] \sinh\left(\dfrac{1}{K(t_0)H_0} \displaystyle\int_{\bar{K}(\bar{t}_1)}^{\bar{K}(\bar{t}_0)} \dfrac{d\bar{K}}{\left(\Omega_{\rho o}\bar{K} + \Omega_x \bar{K}^2 + \Omega_\lambda \bar{K}^4\right)^{1/2}}\right), & k = -1. \end{cases} \tag{9.70}
$$

To obtain a distance redshift relation, we change the $\bar{K}$ integration to an integration over the redshift parameter z, using the relation $\bar{K} = 1/(1+z)$, Eq. (9.29), to obtain

$$
r_1 = \begin{cases} \sin\left(\dfrac{1}{H_0 K(t_0)} \displaystyle\int_0^{z_1} \dfrac{dz}{\left(\Omega_{\rho o}(z+1)^3 + \Omega_x(z+1)^2 + \Omega_\lambda\right)^{1/2}}\right), & k = +1 \\[20pt] \dfrac{1}{H_0 K(t_0)} \displaystyle\int_0^{z_1} \dfrac{dz}{\left(\Omega_{\rho o}(z+1)^3 + \Omega_\lambda\right)^{1/2}}, & k = 0 \\[20pt] \sinh\left(\dfrac{1}{H_0 K(t_0)} \displaystyle\int_0^{z_1} \dfrac{dz}{\left(\Omega_{\rho o}(z+1)^3 + \Omega_x(z+1)^2 + \Omega_\lambda\right)^{1/2}}\right), & k = -1 . \end{cases} \tag{9.71}
$$

The integrals can be performed, at least numerically, to give $r_1(z_1)$. Finally, with these results and the relation $\bar{K} = 1/(1+z)$, Eq. (9.35) can be used to obtain $d_L(z_1)$. In Figure 9.2, $H_0 d_L(z)$ is plotted for model universes with the representative values of $\Omega's$ and q_0 as listed in

Table 9.1. Model Universes

	Ω_ρ	Ω_x	Ω_λ	q_0
A	2	-1	0	1
B	1	0	0	1/2
C	1/3	2/3	0	1/6
D	1/3	0	2/3	$-1/2$

Table 9.1. Note that the curves for universes C and D cross at about $z = 4$. The luminosity distance is the same for both curves for this

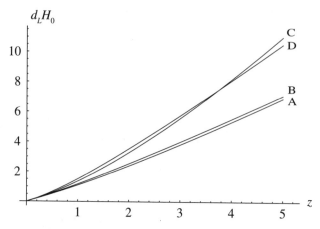

Figure 9.2. Luminosity distance versus redshift.

value of z. One might appreciate from these curves how difficult it is to distinguish universes by measuring the luminosity redshift relation.

9.4.3 Cosmological Dynamics with $\lambda = 0$

The presence of a cosmological term has a significant effect on the development of the universe during an epoch when $|\lambda|$ is greater than or of the order of $8\pi\tilde{G}\rho$ or $8\pi\tilde{G}p$ (see Eqs. (9.50) and (9.51)). We assume in this section that $\lambda = 0$, and in the following section the effect of a nonvanishing cosmological constant is considered. With $\lambda = 0$, if $(\rho(\bar{t}) + 3p(\bar{t}))$ is positive, then Eq. (9.61) implies the rate of change of the slope of the curve K versus t is negative; the curve is concave downward. Furthermore, since at the present time $\dot{\bar{K}} = 1$, the slope at the present time is positive. Together these two properties of the curve imply that at some time $\bar{t}_b$ in the past, $\bar{K}(\bar{t}_b) = 0$. As $\bar{K}(\bar{t}_b) \to 0$ the proper distance $d(t,r)$ approaches zero (Eq. (9.11)). The cosmological metric becomes singular with the energy density becoming infinite. This singular event is sometimes referred to as the "big bang," and $\bar{t}_0 - \bar{t}_b$ would be considered the age of the universe in Hubble time units. Note that if $\ddot{\bar{K}}(\bar{t}) \equiv 0$ in the past (i.e., for $\bar{t} < \bar{t}_0$), then $\bar{t}_0 - \bar{t}_b = 1$; the age of the universe would be the Hubble time. Furthermore, since $\ddot{\bar{K}}(\bar{t}) < 0$,

$$\bar{t}_b - \bar{t}_0 < 1. \tag{9.72}$$

That is, the age of the universe is less than the Hubble time.

In addition to postdicting the beginning of the universe, referred to as the big bang, we can predict possible behaviors of the universe in the future. Of course, such predictions cannot be checked and thus are of less interest.

As noted above, at the present time $p \approx 0$; that is, the equation of state of the universe is that of dust. The universe is expanding and thus will remain dustlike in the immediate future. We will first consider the universe during this expansion phase.

For a dust-filled universe with $\lambda = 0$, Eq. (9.69) becomes

$$\dot{\bar{K}}^2 = \Omega_{\rho_0} \bar{K}^{-1} + \Omega_k. \tag{9.73}$$

By use of this equation we can infer how the fate of the universe depends on the value of k.

1. $k = 0$ ($\Omega_k = 0$): For this case, from Eq. (9.73), we see that $\dot{\bar{K}}$ will not vanish in the future. Thus, the universe will remain dustlike and Eq. (9.73) will remain valid. The solution to this equation is

$$\bar{K}^{3/2}(\bar{t}) = \bar{K}_0^{3/2} + \frac{3}{2} \Omega_{\rho_0}^{1/2} (\bar{t} - \bar{t}_0).$$

From this we see that the "size" of the universe increases without limit.

2. $k = -1$ ($\Omega_k > 0$): Again $\dot{\bar{K}}$ will not vanish in the future. Thus, the universe will remain dustlike, Eq. (9.73) will remain valid, and $\bar{K}$ will continue to increase. Eventually the Ω_k term will dominate the right-hand side of Eq. (9.73). Thus, as $\bar{t} \to \infty, \bar{K}$ increases without bound as $\bar{t}$.

3. $k = +1$ ($\Omega_k < 0$): At some time in the future, $\dot{\bar{K}}$ will vanish and become negative. The density will begin to increase and eventually the universe will cease being dust-like. However, with $(\rho(\bar{t}) + 3p(\bar{t}))$ being positive, the rate of change of the slope of the curve $\bar{K}$ versus $\bar{t}$ is negative, the curve is concave downward, and eventually $\bar{K}$ vanishes—the universe ends.

Though we would not expect the universe to be one of dust back to the big bang and, for the case $k = 1$, forward to the end of the universe, it is instructive to see how a universe of dust would behave. In Figure 9.3 the solution $\bar{K}(\bar{t})$ of Eq. (9.73) is graphed for three universes (A, B, and C of Table 9.1), for which $k = 1, 0, -1$, respectively.

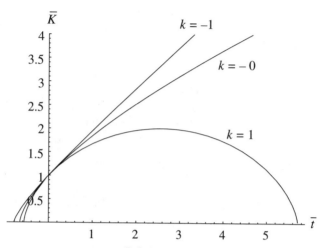

Figure 9.3. $\bar{K}(\bar{t})$ for a universe of dust.

During the epoch for which $p \approx 0$, referred to as the *matter domi-nated era*, as K becomes small, ρ becomes large. Eq. (9.68) merely reflects the fact that the mass contained in a coexpanding volume is fixed. We might expect that as ρ grows back in time, the equation of state would gradually switch over to that of a relativistic hot gas, namely,

$$\rho = 3p. \tag{9.74}$$

The epoch, for which Eq. (9.74) is a good approximation, is called the *radiation-dominated era*. In that case, Eq. (9.58) leads to

$$\rho \propto \bar{K}^{-4}. \tag{9.75}$$

Using this in Eq. (9.50) we see that, at a time when $K(t)$ is small, for instance, early in the life of the universe, then

$$\dot{\bar{K}}(\bar{t}) \propto \bar{K}^{-1}, \tag{9.76}$$

and thus

$$\bar{K} \propto \bar{t}^{1/2}. \tag{9.77}$$

The condition for a particle horizon exists. (See Eq. (9.15).) In fact, even if the matter-dominated era obtained back to the big bang, a particle horizon would still exist.

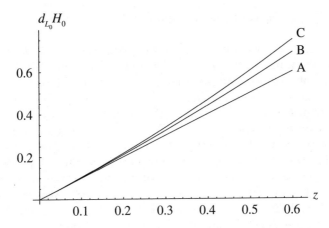

Figure 9.4. Luminosity distance versus redshift for universes A, B, and C.

In a universe with $\lambda = 0$, from Eqs. (9.64) and (9.65),

$$\frac{\rho_0}{\rho_c} = 2q_0. \tag{9.78}$$

Thus, we have the following possibilities:

$$
\begin{aligned}
k &= 1 \quad q_0 > 1/2 \\
k &= 0 \quad q_0 = 1/2 \\
k &= -1 \quad q_0 < 1/2.
\end{aligned}
\tag{9.79}
$$

A direct observational determination of q_0 is obtained by measuring how the luminosity distance d_L varies with z, the redshift parameter. Figure 9.4, which graphs Eq. (9.42), shows the deviation of $d_L H_0$ from linear for values of universes A, B, and C of Table 9.1, which have $q_0 = 1, 1/2, 1/6$, respectively. We see then that a measurement of H_0 determines the critical density ρ_c, whereas a measurement of both H_0 and the deceleration parameter q_0 determines the present density of the universe ρ_0. Conversely, a measurement of H_0 and a determination of ρ_0 determines q_0. An independent determination of all three is a consistency check on the theory. Unfortunately, the reasonable assumption that essentially all of the matter of the universe is concentrated in the visible mass of galaxies gives rise to discrepancies in the dynamics of galaxies and clusters of galaxies, perhaps indicating the presence of a large amount of "dark matter." The amount and nature of this dark matter are among the outstanding questions facing present-day cosmology.

9.4.4   Cosmological Dynamics with $\lambda \neq 0$

Admitting the possibility of a nonvanishing cosmological constant widens the range of the possible dynamics of the universe. As noted before, observations imply that $|\lambda|$ is very small; $|\lambda| < 10^{-46} m^2$. Nevertheless, an even smaller $|\lambda|$ can have a significant effect on cosmological dynamics. If $|\Omega_\lambda| \approx \Omega_{\rho_0}$ then $|\lambda| \approx 4\pi \tilde{G} \rho_0 = 4\pi G \rho_0 /c^2$. Since $\rho_0 \approx 10^{-26} Kg/m^3$, $|\lambda| \approx 10^{-54} m^2$. A universe even with $|\Omega_\lambda|$ comparable to Ω_{ρ_0} has a very small λ. One sure indication of a nonvanishing λ would be the observation of a negative value of q_0. (See Eq. (9.65).) Evidence from observations of supernovae at up to a redshift of 0.7 indicates that in fact $q_0 < 0$ (Riess et al. 1998). Also, it is generally believed that the universe is very nearly flat at the present time, or $\Omega_k \approx 0$ (see Sec. 9.5), and thus Eq. (9.63) becomes

$$1 \approx \Omega_{\rho_0} + \Omega_\lambda. \tag{9.80}$$

Furthermore, the size of cosmic matter density, including the so-called dark matter inferred from galaxy dynamics, seems to imply that $\Omega_{\rho_0} < .5$. So, when $\Omega_k \approx 0$, then $\Omega_\lambda > .5$.[3] Perhaps the best fit to data is $\Omega_{\rho_0} \approx 1/3$ and thus $\Omega_\lambda \approx 2/3$. Consider then, as a model of the universe, a dust-filled universe with $\Omega_k = 0$, $\Omega_{\rho_0} = 1/3$ and $\Omega_\lambda = 2/3$. This is model universe D of Table 9.1. Then Eq. (9.69) becomes

$$\dot{K}^2 = \frac{1}{3} \bar{K}^{-1} + \frac{2}{3} \bar{K}^2. \tag{9.81}$$

This equation can be integrated with the boundary condition $\bar{K}(\bar{t}_0 = 0) = 1$. The resulting solution is plotted in Figure 9.5. The solution with $\Omega_{\rho_0} = 1$, corresponding to model universe C of Table 9.1, is also included in the figure for comparison. Notice that the age of the universe for $\Omega = 2/3$ is approximately $.94 H_0^{-1}$, whereas for the flat universe, with $\lambda = 0$, it is smaller, about $.65 H_0^{-1}$. This is not surprising, since the $\lambda = 0$ universe has much more mass for the same Hubble constant. Universe D has $q_0 = -1/2$. (See Eq. (9.64).)

In Figure 9.6, Eq. (9.42), the variation of the luminosity-distance with the redshift, is graphed for universe D. In the same figure, the graph for a universe with $\lambda = 0$ and $q_0 = 1/6$, that is, universe C, is shown to give an idea of the measurements required to distinguish such cases.

---

[3] See Section 9.5.2 for an argument as to why Ω_k is approximately zero.

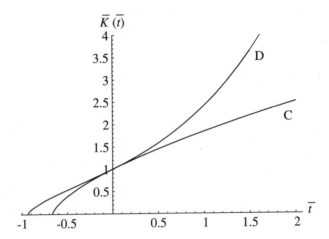

Figure 9.5. $\bar{K}(\bar{t})$ for universes C and D.

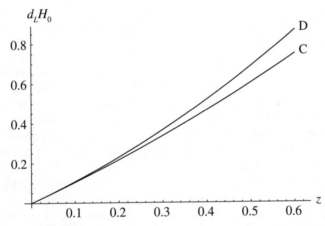

Figure 9.6. Luminosity distance versus redshift for universes C and D.

9.5 The Early Universe

The gross behavior of the universe at early times depended on the equation of state of the ideal fluid at the expected high densities and the expected associated high temperatures. We will not discuss this in any detail as it would take us far afield into the discussion of the thermodynamics of the interaction of elementary particles and their descendants at these high densities and temperature. However, observations such as the relative abundance of nuclei and even the size of the clustering of galaxies give the cosmologist insight into the correctness of the developing picture of the universe—or raise questions about the consistency of the picture.

9.5.1 The Cosmic Microwave Background Radiation

The strongest evidence for the existence of this high density–high temperature phase of the universe is the thermal *cosmic background radiation* (CMB). At some early time, we expect that the electromagnetic radiation was in thermodynamic equilibrium with matter consisting of electrons and nuclei, such as protons. This electromagnetic radiation would be black body in character. The energy density per frequency interval $\mu(\nu, T)$, which is characteristic of the temperature, is given by the Planck formula,

$$\mu(\nu, T) = \frac{8\pi h \nu^3}{\exp(h\nu/kT) - 1}. \tag{9.82}$$

As the universe expanded and cooled, the radiation would remain in thermal equilibrium with matter by ionizing the atoms as they were formed. This scenario would continue until the temperature dropped to the point that the energy of a typical photon was too small to ionize atoms—the radiation would thermally decouple from the matter. This would occur at a temperature of about 4000 K.[4] Subsequent to this decoupling, the expansion of the universe causes the radiation to undergo a redshift, which, with the concomitant expansion of a volume element, maintains the black body character of the radiation with a lower effective temperature. That the black body character is maintained is not too difficult to derive. In addition, the relation among the decoupling temperature, the present effective temperature, the size of the universe at decoupling and the present size is easily obtained.

In the following we denote the decoupling time as t_d, the decoupling temperature as T_d, the present time as t_0, and the present effective temperature of the radiation as T_0. It follows from the definition of $\mu(\nu)$ and the quantum condition $E = h\nu$ that the number of photons in a volume element dV in a range of frequencies $d\nu$ is given by $\mu/h\nu$. With the assumption that the decoupled photons are not absorbed, the number of photons of a fixed direction in a frequency

---

[4] One would not expect a sharp decoupling. In fact, the high-frequency tail of the black-body radiation would keep the radiation in thermal contact down to a lower temperature. Thus, the effective decoupling temperature would be somewhat lower then 4000 K.

range $d\nu_d$ and a propagating volume element dV_d at the time of decoupling is equal to the number in a redshifted frequency range $d\nu_0$ and expanded volume element dV_0 at the present time. Thus,

$$\frac{8\pi\nu_d^2}{\exp(h\nu_d/kT_d)-1}\,d\nu_d\,dV_d = \frac{\mu(\nu_0)}{h\nu_0}\,d\nu_0\,dV_0. \qquad (9.83)$$

The photons, in a frequency range $d\nu_d$ at decoupling, would be redshifted to be in a frequency range $d\nu_0$ at the present time, given by (see Eq (9.26))

$$d\nu_0 = \frac{K(t_d)}{K(t_0)}\,d\nu_d, \qquad (9.84)$$

and the expanded volume element dV_0 is related to the volume element dV_d by

$$dV_0 = \left(\frac{K(t_0)}{K(t_d)}\right)^3 dV_d. \qquad (9.85)$$

Using Eqs. (9.84) and (9.85) in Eq. (9.83), we find that

$$\mu(\nu_0) = \frac{8\pi h\nu_0^3}{\exp(h\nu_0 K(t_0)/kT_d K(t_d))-1}. \qquad (9.86)$$

Thus, the decoupled radiation maintains its black-body character with a present *effective* temperature T_0 given by

$$T(t_0) = \frac{K(t_d)}{K(t_0)}\,T(t_d). \qquad (9.87)$$

The observed value of $T(t_0)$ is about $2.7K$, which corresponds to a ratio

$$\frac{K(t_d)}{K(t_0)} \approx 10^{-3}. \qquad (9.88)$$

Recently, beautiful and sensitive measurements have been made of this CMB radiation. The radiation has been observed in different directions with surprisingly uniform results, with the effective temperature anisotropy, $\Delta T/T$, being on the order of 10^{-6}, after correction for the Doppler shift of the motion of our solar system in the universe (Bennet et al. 1996).

9.5.2   Inflation

Why is it surprising that the thermal background radiation is so uniformly the same when observed in different directions? Consider

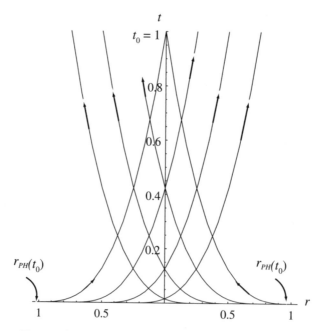

Figure 9.7. World lines of light emitted at the "big bang."

the observations of radiation coming from one direction, and then from the opposite direction. The radiation from one direction (say $\phi = 0$) was emitted at a radial position r_d given by

$$\int_{t_d}^{t_0} \frac{dt}{K(t)} = \int_0^{r_d} \frac{dr}{(1 - kr^2)^{1/2}}. \tag{9.89}$$

Here, as before, t_d is the cosmic time at which decoupling occurred. The radiation from the opposite direction (say, $\phi = \pi$) was emitted at the same radial position. But these are separated positions, and the question arises as to why they should have the same temperature to such a high degree of accuracy. The reader may of course say that the cosmological principle demands this. But should the isotropy of the universe at a given time hold over distances between points that were never in causal contact? This would seem to indicate that the material at these two positions were in thermal equilibrium at the (cosmic) time of decoupling, which in turn would seem to imply that one position was within the particle horizon (see Eq. (9.14)) of the other at the decoupling time.

We can understand what is involved by studying Figure 9.7. The figure depicts world lines of photons that were emitted toward us at time zero at various r and from opposite directions, say, $\theta = 0$ and $\theta = \pi$. The world lines are calculated for the same case as that of Figure 9.1, that is, for a universe with $k = 0, \lambda = 0$, and dust filled with the pressure assumed to be zero all the way back to $t = 0$. The set of four world lines for $\theta = 0$, originate at $t = 0$ at $r = 1, 3/4, 1/2$, and $1/4$; similarly, for the $\theta = \pi$ world lines. We can read off the figure approximately when a position of a given r with $\theta = 0$ comes within the horizon of the position r with $\theta = \pi$, that is, when they come in causal contact. For example, the two positions for $r \approx .1$ clearly are in causal contact at $t = .7$, the time at which the light emitted at $r \approx .1$ reaches us. In fact, they are in causal contact before $t = .10$. The two positions $r \approx .3$ have just come in causal contact at $t \approx .3$, the time at which the light emitted at $r \approx .3$ reaches us, whereas the two positions $r \approx .5$ are clearly not in causal contact at $t = .15$, the time at which the light emitted at $r \approx .5$ reaches us.

We see that the time t_r^c at which the position $r, \theta = \pi$ comes in causal contact with $r, \theta = 0$ is given by

$$\int_0^{t_r^c} \frac{dt}{K(t)} = \int_{-r}^r \frac{dr}{(1 - kr^2)^{1/2}} = 2 \int_0^r \frac{dr}{(1 - kr^2)^{1/2}}. \tag{9.90}$$

From this, the condition that the positions $r_d, \theta = \pi$ and $r_d, \theta = 0$ are in causal contact at decoupling time t_d is

$$\int_0^{t_d} \frac{dt}{K(t)} > 2 \int_0^{r_d} \frac{dr}{(1 - kr^2)^{1/2}} = 2 \int_{t_d}^{t_0} \frac{dt}{K(t)}. \tag{9.91}$$

The universe has undergone a large expansion by a factor of 10^3 since decoupling. The validity of this inequality is determined by how the expansion from $t = 0$ to $t = t_d$ compares with this. We first change the t integration to a K integration. Eq. (9.91) becomes

$$\int_0^{K(t_d)} \frac{dK}{\dot{K}K} > 2 \int_{K(t_d)}^{K(t_0)} \frac{dK}{\dot{K}K}. \tag{9.92}$$

We consider a universe for which $k = \lambda = 0$. Assume that the integrals are dominated by the contribution during the matter-dominated era for which

$$\dot{K}K = H_0 K(t_0)^{3/2} K^{1/2} = CK^{1/2}. \tag{9.93}$$

Here we have used Eq. (9.69), with $\Omega_k = \Omega_\lambda = 0$, and have set $H_0 K(t_0)^{3/2} = C$. The required inequality, Eq. (9.92), becomes

$$\frac{K(t_d)^{1/2}}{2C} > \frac{K(t_0)^{1/2} - K(t_d)^{1/2}}{C} \tag{9.94}$$

or

$$\frac{1}{2}\frac{K(t_d)^{1/2}}{K(t_0)^{1/2}} > 1 - \frac{K(t_d)^{1/2}}{K(t_0)^{1/2}}. \tag{9.95}$$

Clearly, this inequality is not satisfied since $K(t_d)/K(t_0) \approx 10^{-3}$. We can also see this from Figure 9.7, since the figure is drawn for a matter-dominated universe extending back to $t = 0$. For such a universe $K \propto t^{2/3}$ and thus $t_d \approx 3 \times 10^{-5}$. But we saw that $t \approx .3$ is the earliest time that light, emitted from opposite directions and just now reaching us, came from causally connected sources. One might argue that at least for the left integral of Eq. (9.92), a large part of the contribution comes from the radiation-dominated era, that is, for which $\rho \propto K^{-4}$ and $\dot{K}K \approx$ constant. This makes the situation worse. The contribution to the integral during the radiation-dominant epoch becomes proportional to $K(t_r)$, the value at which the radiation epoch ends. Note that during a radiation-dominated era $K \propto t^{1/2}$, whereas for the matter-dominated era, $K \propto t^{2/3}$, and a faster expansion results. If the inequality is to be satisfied, an expansion much faster than that of the matter-dominated era must occur for some "inflationary" period before decoupling. An example of such a period would be one in which the expansion is exponential, to whit,

$$K \propto \exp(at), \tag{9.96}$$

with a positive constant a. K would then satisfy the equation

$$\dot{K}K = aK^2. \tag{9.97}$$

For such a period beginning at t_b and ending at t_e, before decoupling, the contribution to the left-hand side of Eq. (9.90) is

$$\int_{K(t_b)}^{K(t_e)} \frac{dK}{aK^2} = \frac{K(t_e) - K(t_b)}{aK(t_e)K(t_b)}. \tag{9.98}$$

If there is a large expansion during this period, that is, $K(t_e) >>> K(t_b)$, then the inequality is surely satisfied if

$$\frac{C}{4aK(t_0)^{1/2}K(t_b)} = \left(\frac{H_0}{4a}\right)\frac{K(t_0)}{K(t_b)} > 1 - \frac{K(t_d)^{1/2}}{K(t_0)^{1/2}}. \tag{9.99}$$

If the inflationary expansion is large enough, this will hold. We can see the effect of the inflationary expansion period on the world lines of Figure 9.7. During the period that K satisfies Eq. (9.96) $r \propto \exp(-at)$, t approaches zero logarithmically with increasing r. The world lines reach much farther out in r.

As noted by Guth (1981), existence of a period that gives rise to an exponential expansion is suggested by quantum theory models that unify the weak electromagnetic and strong interactions and that predict a phase transition at high temperatures at which the system might attain a supercooled state whose vacuum would contribute to the energy-momentum tensor a term $\alpha g_{\alpha\beta}$, with α assuming a large positive value for some time. During this time, the effect is the same as that of a large positive cosmological constant of value $\lambda = 8\pi\tilde{G}\alpha$. For a such an effective cosmological term, eventually the constant λ term dominates the k and ρ terms of Eq. (9.50), and Eq. (9.96) results with $a = (8\pi\tilde{G}\alpha/3)^{1/2}$.

A concomitant effect of the standard inflationary model is that, after inflation, that is, after the universe condenses from its supercooled state, the universe is very nearly flat. Thus, $|\Omega_k|$ is much smaller than $\Omega_{\rho 0}$. Note that our model universes B and D of Table 9.1 satisfy this condition.

After condensation occurs, the universe would be in the very high density and high temperature condition of the big bang.

9.5.3 Cosmic Microwave Background and Cosmological Parameters

We have seen that the determination of the cosmological parameters, the Ω's, is a primary goal in cosmological studies. And, as we have seen, a determination of the redshift-distance relation contributes to the measurement of these parameters. Surprisingly, the study of the very small anisotropy of the CMB is proving to be a marvelous second tool in this study. As an illustration of the effect of geometry on the anisotropy of the CMB, we will discuss how geometry affects the position of the first "acoustic" peak. Imagine a small density perturbation present at the big bang or, equivalently, immediately after the universe condenses following the inflationary expansion. Such a perturbation will expand with the velocity of sound c_s in the hot, dense relativistic fluid to the time of decoupling t_d out to a (radial) position r_{SH}, the "sound horizon," related by

$$\int_0^{r_{SH}} \frac{dr}{(1-kr^2)^{1/2}} = c_s \int_0^{t_d} \frac{dt}{K(t)}. \qquad (9.100)$$

The proper size d_{SH}, Eq. (9.11), of this sound horizon at the time of decoupling is given by

$$d_{SH} = K(t_d) \int_0^{r_{SH}} \frac{dr}{(1-kr^2)^{1/2}} = c_s K(t_d) \int_0^{t_d} \frac{dt}{K(t)}$$

$$= \frac{c_s K(t_d)}{K(t_0) H_0} \int_0^{\bar{K}(t_d)} \frac{d\bar{K}}{\dot{\bar{K}}\bar{K}}. \tag{9.101}$$

We have again changed the t integration to a $\bar{K}$ integration. Recall that $\bar{K}(t_d) \approx 10^{-3}$. From Eq. (9.60), the denominator of the last integral is

$$\dot{\bar{K}}\bar{K} = (\Omega_{\rho 0} \frac{\rho(\bar{t})}{\rho_0} \bar{K}^4 + \Omega_k \bar{K}^2 + \Omega_\lambda \bar{K}^4)^{1/2}. \tag{9.102}$$

With the assumptions that the relativistic equation of state obtains from decoupling back to the big bang and a zero-pressure dust-filled universe back to decoupling, the energy density in this equation can be written as

$$\rho(\bar{t}) = \rho(t_d) \frac{\bar{K}(\bar{t}_d)^4}{\bar{K}(\bar{t})^4} = \frac{\rho_0}{\bar{K}(\bar{t}_d)^3} \frac{\bar{K}(\bar{t}_d)^4}{\bar{K}(\bar{t})^4} = \rho_0 \frac{\bar{K}(\bar{t}_d)}{\bar{K}(\bar{t})^4}.$$

Here, the first equality results from applying Eq. (9.75), which is true for a relativistic fluid, and the second from Eq. (9.68), which is valid for a dust-filled universe. With this expression for ρ substituted into Eq. (9.102) and the result substituted into Eq. (9.101), we obtain

$$d_{SH} = \frac{c_s K(t_d)}{K(t_0) H_0} \int_0^{\bar{K}(t_d)} \frac{d\bar{K}}{(\Omega_{\rho 0} \bar{K}(t_d) + \Omega_k \bar{K}^2 + \Omega_\lambda \bar{K}^4)^{1/2}} \approx \frac{c_s K(t_d) \bar{K}(t_d)^{1/2}}{K(t_0) H_0 \Omega_\rho^{1/2}} \tag{9.103}$$

The proper size of the sound horizon is a "perpendicular" proper size as viewed by us. Thus, the angular size of this sound horizon $\delta\theta_{SH}$, as viewed by us, is related to this proper size d_{SH} by (see Eq. (9.13))

$$d_{SH} = K(t_d) r_d \delta\theta_{SH}, \tag{9.104}$$

with

$$r_d = \begin{cases} \sin\left(\frac{1}{K(t_0) H_0} \int_{\bar{K}(t_d)}^1 \frac{d\bar{K}}{(\Omega_{\rho 0} \bar{K} + \Omega_x \bar{K}^2 + \Omega_\lambda \bar{K}^4)^{1/2}}\right), & k = +1 \\ \frac{1}{K(t_0) H_0} \int_{\bar{K}(t_d)}^1 \frac{d\bar{K}}{(\Omega_{\rho 0} \bar{K} + \Omega_\lambda \bar{K}^4)^{1/2}}, & k = 0 \\ \sinh\left(\frac{1}{K(t_0) H_0} \int_{\bar{K}(t_d)}^1 \frac{d\bar{K}}{(\Omega_{\rho 0} \bar{K} + \Omega_x \bar{K}^2 + \Omega_\lambda \bar{K}^4)^{1/2}}\right), & k = -1. \end{cases} \tag{9.105}$$

With r_d known, Eqs. (9.103) and (9.104) determine the angular size of the sound horizon:

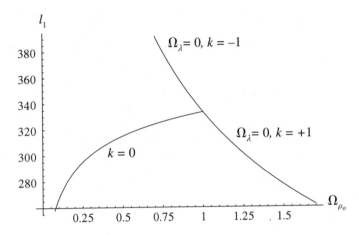

Figure 9.8. First acoustic peak, l_1 vs. Ω_{ρ_0} for three sets of Ω's.

$$\delta\theta_{SH} = \frac{c_s \bar{K}(t_d)^{1/2}}{K(t_0) H_0 \Omega_{\rho_0}^{1/2} r_d}. \tag{9.106}$$

But what properties of the CMB are related to $\delta\theta_{SH}$? The sound-waves are density waves and thus are temperature waves. The temperature is higher (lower) than the ambient temperature where the density is larger (smaller) than the background density. Thus, at decoupling the soundwaves leave a temperature variation imprint. These waves have an extent as large as the sound horizon—an angle extent as large as $\delta\theta_{SH}$. Performing a harmonic analysis of the observed temperature variation with angle—that is, expanding the observations in terms of $\cos(\ell\theta)$—one would expect the result would peak at values of ℓ such that an integer number of half-waves would cover the sound horizon. (One should perform a spherical harmonic analysis—we are dealing with a two-dimensional surface.) The acoustic peaks should occur at

$$\ell_j = \frac{j\pi}{\delta\theta_{SH}} = \frac{j\pi K(t_0) H_0 \Omega_{\rho_0}^{1/2} r_d}{c_s \bar{K}(t_d)^{1/2}}. \tag{9.107}$$

The positions of the peaks clearly depend on the geometry through the terms $\Omega_{\rho_0}^{1/2}$ and r_d. The position of the first peak, ℓ_1, calculated using Eq. (9.107), is graphed in Figure 9.8 for three sets of Ω's:(1) $\lambda = 0$, $k = +1$, (2) $\lambda = 0$, $k = -1$, and (3) $k = 0$. $\bar{K}(t_d)$ is taken to be 10^{-3} and $c_s = 3^{-1/2}$.

Recent analysis of the anisotropy of CMB data gives a strong indication for a flat universe with $\Omega_\lambda \approx 2/3$ and $\Omega_{\rho_0} \approx 1/3$, the values of our universe D (Melchorri et al. 2000). As we have seen, for such a universe $q_0 < 0$, which agrees with the result of the high z redshift observations noted before.

Other questions need to be answered, supported by observations, that bear on the consistency of the emerging view of the big bang origin of our universe. And there are apparent difficulties. But it must be said that Einstein's general relativity theory, coupled with the cosmological principle, have been successful to a remarkable extent in developing a coherent picture of our universe.

9.6 Exercises

1. Consider Einstein's field equations with a nonvanishing cosmological constant λ. (a) Show that there exists a static solution only if $k = 1$ and $\lambda > 0$. (b) For such a solution for a dust-filled universe, derive the relations of the "radius" K and the density ρ to the cosmological constant λ.

2. Again, consider Einstein's field equations for a $k = 0, \pm 1$, homogeneous, isotropic, and empty universe (i.e., $\rho = p = 0$) and with a positive cosmological constant λ. (a) What are the resulting space-time metrics? Use your program to calculate (b) the space-time curvature tensors $R_{\alpha\beta\sigma\rho}$ and (c) the Ricci scalars for these metrics. (d) Show that these metrics are *space-time* homogeneous and isotropic. (Since all have the same constant Ricci scalar, they are equivalent—they are different forms of the *de Sitter* metric.)

3. For the metric of Eq. (8.65), which is called the anti–de Sitter metric, calculate (a) the space-time curvature tensors $R_{\alpha\beta\sigma\rho}$ and (c) the Ricci scalars. (d) Show that this metric is *space-time* homogeneous and isotropic. (e) Show that this metric satisfies Einstein's field equations for an empty universe with a negative cosmological constant.

4. Assume that after decoupling the photon gas maintains thermal equilibrium with itself. Using the knowledge that, for such a photon gas, $\rho \propto T^4$, show that $T \propto K$, and thus Eq. (9.86) is valid.

Suggested Additional Reading

For a very readable introduction to special relativity and a source of many exercises, see Taylor and Wheeler (1966).

I just touched the surface of electromagnetic theory, using it as an example of a relativistic field. An extensive treatment of the relativistic formulation of electromagnetic theory is given in the classic book by J. D. Jackson (1975). For accessible discussions of relativistic fluids and associated energy-momentum tensors, see Schutz (1985) and Weinberg (1972).

Our treatment of tensor analysis is based on coordinate transformations. Students wishing to study the more modern coordinate-free approach to differential geometry applied to general relativity are referred to Wald (1984). The mathematically inclined student might want to look at Nash and Sen (1983) for an introduction to mathematical constructs, such as manifolds, forms, connections, etc., used in general relativity.

In the section on equilibrium stellar interiors, I did not cover reasonable equations of state for high density nor did I cover the evolution of stellar interiors. These topics must be studied before questions concerning the formation of black holes via stellar evolution can be answered. A popular account of the science and history of black holes is given in Thorne (1994).

Topics at the interface of astrophysics and cosmology are discussed in marvelous detail in Peebles (1993). Astrophysical observations having a defining effect on cosmological studies are being made at such a fast pace that many new results are not in older reviews. A good source for semitechnical articles on recent advances is the magazine *Scientific American*.

References

Bennet, C. et al. (1996). *Astrophysical Journal* 464:1.

Bernstein, J. (1973). *Einstein,* New York: Viking Press.

Birkhoff, G. (1923). *Relativity and Modern Physics.* Cambridge, Mass.: Harvard University Press.

Buchdahl, H. A. (1959). *Physical Review* 116:1027.

Dyson, F., Eddington, A., and Davidson, C. (1920). *Phil. Trans. Roy. Soc.* 220:291.

Einstein, A. (1916). *Annalen der Physik* 49:769.

Einstein, A., Lorentz, H., and Minkowski, H. (1923). *Principle of Relativity.* New york: Dover.

Freedman, W., et al. (2001). *Astrophysical Journal* 553:47.

Guth, A. (1981). *Physical Review D* 23:347.

Harrison, E. R. (1987). *Darkness at Night: A Riddle of the Universe.* Cambridge Mass.: Harvard University Press.

Hubble, E. (1929). *Proc. Nat. Acad. Sci.* 15:168.

Jackson, J.D. (1975). *Classical Electrodynamics.* New York: Wiley

Melchorri, A., et al. (2000). *Astrophysical Journal Letters* 536:63.

Nash, C., and Sen, S. (1983). *Topology and Geometry for Physicists.* London: Academic Press.

Peebles, P.J.E. (1993). *Principles of Physical Cosmology.* Princeton, N.J.:Princeton University Press.

Pound, R. V., and Rebka, G.A. (1960). *Physical Review Letters* 4:337.

Riess, A., et al. (1998). *Astronomical Journal* 116(3):1009.

Rindler, W. (1956). *Mon. Not. R. Astron. Soc.* 116:662.

Rindler, W. (1991). *Introduction to Special Relativity.* 2d ed. New York: Oxford University Press.

Schilpp, P.A., ed.(1949). *Albert Einstein: Philosopher-Scientist.* Evanston, Ill.: Library of Living Philosophers.

Schutz, B.F. (1985). *A First Course in General Relativity*. Cambridge, U.K.: Cambridge University Press.

Thorne, K. S. (1994). *Black Holes and Time Warps*. New York: Norton.

Taylor, E.F., and Wheeler, J. A. (1966) *Spacetime Physics*. San Francisco; Freeman.

Wald, R. M. (1984). *General Relativity*, Chicago: University of Chicago Press.

Weinberg, S. (1972). *Gravitation and Cosmology*. New York: Wiley.

Index